复旦卓越·电工电子系列

TECHNICAL APPLICATION OF
INDUSTRIAL FREQUENCY CONVERTERS

工业变频器技术应用教程

白 锐 王贺彬 赵越岭 吴 静 编著

复旦大学出版社

内容提要

本书共5章，前3章为设备篇，后2章为实验实训篇。设备篇包括罗克韦尔1336 Impact变频器、PowerFlex系列变频器介绍，以及罗克韦尔的网络构架与网络基础。实验实训篇包括1336 Impact变频器与PowerFlex系列变频器的实验原理和实验过程。第1章介绍1336 Impact变频器的结构特性、功能特点、接口与通信、参数与设置、故障检修和相关应用。第2章讲述PowerFlex系列变频器的功能和类型特点，以及PowerFlex 4/40/400 系列变频器、PowerFlex 520 系列变频器和PowerFlex 750系列变频器的功能特点与应用环境。第3章讨论罗克韦尔的网络体系与集成架构特点，工业以太网(EtherNet/IP)的理论要点、网络协议、网络模型和技术应用，DeviceNet总线的技术特点、网络接口和对象模型，ControlNet总线的理论要点、技术特点与应用。第4章包括1336 Impact变频器面板操作控制实验、控制端子应用实验、单极性/双极性的模拟量输入实验和多段速运行操作实验。第5章设计了基于DeviceNet的网络配置及通信、控制电动机启停和变频调速的实验，以及基于ControlNet的网络配置、通信及应用和变频器频率控制的实验。

前　言

随着工业变频器的迅速发展,工业变频技术在工业现场发挥的作用也变得更加明显。目前在高等学校、高职院校中关于变频技术的介绍都过于片面,现有的关于变频技术与变频器的教材或书籍存在一定的局限性。从培养学生工程实践能力的角度来看,目前教材或书籍并不适用于学生动手实践的教学。无论是基本原理、结构架构,还是生产厂家、选型配置,目前教材都局限于理论层面,缺少动手实践部分的深度和广度。考虑到变频技术的应用性较强,为配合理论课教学,我们编写了本应用教程。本教程基于罗克韦尔(Rockwell)网络架构与罗克韦尔变频器产品设计了相关实验例程,学生通过这些实验例程可以更好地掌握交流调速技术和工业控制网络的知识要点。

本书结合党的二十大精神,按照"实施人才强国战略"的要求,融合专业特点进行编写。在编写中充分体现新时代中国特色社会主义建设伟大事业的人才培养要求和科教兴国战略,致力于培养德才兼备的高素质人才。整个应用教程共分5章,前3章为设备篇,后2章为实验实训篇。设备篇包括罗克韦尔1336 Impact 变频器、PowerFlex 系列变频器介绍,以及罗克韦尔的网络构架与网络基础。实验实训篇包括1336 Impact 变频器与 PowerFlex 系列变频器的实验原理和实验过程。第1章介绍1336 Impact 变频器的结构特性、功能特点、接口与通信、参数与设置、故障检修和相关应用。第2章讲述 PowerFlex 系列变频器的功能和类型特点,以及 PowerFlex 4/40/400 系列变频器、PowerFlex 520 系列变频器和 PowerFlex 750 系列变频器的功能特点与应用环境。第3章讨论罗克韦尔的网络体系与集成架构特点,工业以太网(EtherNet/IP)的理论要点、网络协议、网络模型和技术应用,DeviceNet 总线的技术特点、网络接口和对象模型,ControlNet 总线的理论要点、技术特点与应用。第4章包括1336 Impact 变频器面板操作控制实验、控制端子应用实验、单极性/双极性的模拟量输入实验和多段速运行操作实验。第5章设计了基于 DeviceNet 的网络配置及通信、控制电动机启停和变频调速的实验,以及基于 ControlNet 的网络配置、通信及应用和变频器频率控制的实验。

本书是辽宁工业大学的立项教材,并由辽宁工业大学资助出版。在本教程的编写中,白锐负责本书的统筹规划并编写了第1章和第2章,王贺彬编写了第4章,赵越岭编写了第5章,吴静编写了第3章。本教程中的实验得到辽宁工业大学罗克韦尔工业控制网络平台的支撑，在此对罗克韦尔公司表示感谢！

　　由于编者水平有限，特别是对 Allen-Bradley 品牌变频器在实际应用中的积累还不够，书中难免出现错误，敬请广大读者批评指正。

编　者

2022 年 12 月

目 录

设 备 篇

第1章 1336 Impact 变频器介绍 (003)
  §1.1 罗克韦尔 1336 Impact 变频器概述 (003)
  §1.2 1336 Impact 变频器的结构特性 (003)
  §1.3 1336 Impact 变频器的功能特点 (004)
  §1.4 1336 Impact 变频器的接口与通信 (005)
  §1.5 1336 Impact 变频器的参数与设置 (009)
  §1.6 1336 Impact 变频器的故障检修 (013)
  §1.7 1336 系列变频器的应用 (014)

第2章 PowerFlex 系列变频器介绍 (016)
  §2.1 PowerFlex 系列变频器的功能特点 (016)
  §2.2 PowerFlex 变频器的类型 (019)
  §2.3 PowerFlex 4/40/400 系列变频器 (023)
  §2.4 PowerFlex 520 系列变频器 (025)
  §2.5 PowerFlex 750 系列变频器 (026)

第3章 罗克韦尔自动化网络体系 (027)
  §3.1 NetLinx 集成架构 (027)
  §3.2 NetLinx 体系结构的优点 (028)
  §3.3 工业以太网理论要点 (029)
  §3.4 DeviceNet 总线理论要点 (034)
  §3.5 ControlNet 总线理论要点 (037)
  §3.6 不同网络的优势与选型 (042)

实验实训篇

第4章　1336 Impact 变频器实验 ………………………………………………………（049）
　　§4.1　1336 Impact 变频器的面板操作控制 ……………………………………（049）
　　§4.2　1336 Impact 变频器的控制端子应用 ……………………………………（053）
　　§4.3　1336 Impact 变频器的模拟信号操作与运行（模拟量的单极性控制）…（058）
　　§4.4　1336 Impact 变频器的模拟信号操作与运行（模拟量的双极性控制）…（063）
　　§4.5　1336 Impact 变频器的多段速运行操作 …………………………………（066）

第5章　PowerFlex 系列变频器实验 …………………………………………………（070）
　　§5.1　基于 DeviceNet 的变频器实验 …………………………………………（070）
　　§5.2　基于 ControlNet 的变频器实验 …………………………………………（090）

附录A　1336 Impact 变频器常用参数 ………………………………………………（107）
附录B　PowerFlex 40 变频器常用参数 ……………………………………………（126）

参考文献 ………………………………………………………………………………（149）

设备篇

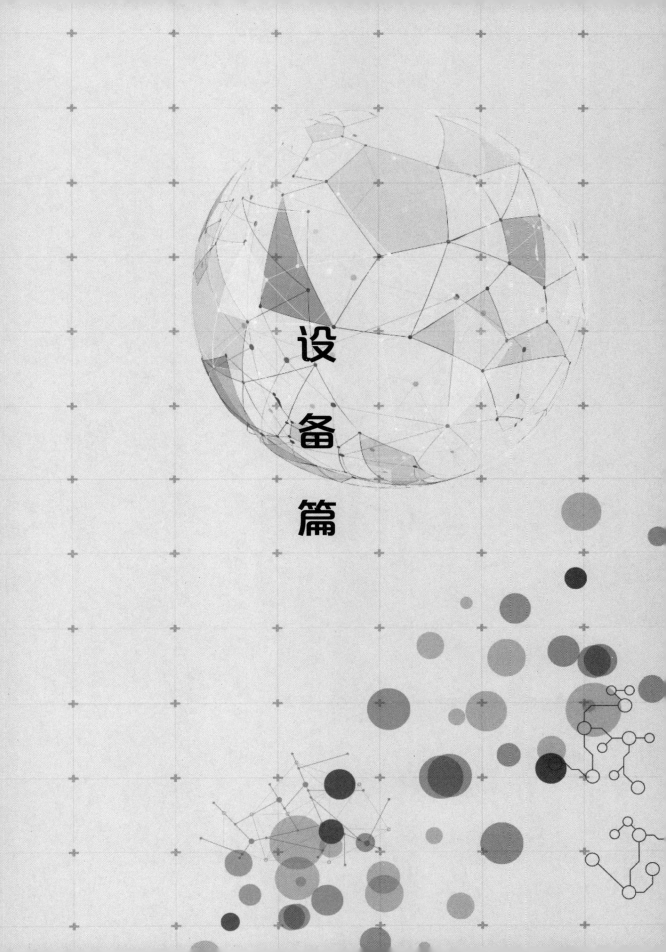

第 1 章
1336 Impact 变频器介绍

§1.1 罗克韦尔 1336 Impact 变频器概述

罗克韦尔 1336 Impact 变频器采用 Allen-Bradley(以下简称 A－B)公司的 FORCE 技术。FORCE 技术是基于磁场定向的控制方法,其优点是实现了电动机磁场和转矩的独立控制,从而满足了电动机速度和转矩的精准控制。同时,1336 Impact 变频器具有快速的电动机整定步骤,操作者可通过变频器面板菜单输入电动机参数,可以对模拟量、数字量和各种通信模块进行组态,满足用户和现场的要求。

1336 Impact 变频器和 1336 PLUS,1336 FORCE 变频器系列具有许多通用件,包括相同的电源结构、通信接口和通信模块,方便其与各种网络连接。

§1.2 1336 Impact 变频器的结构特性

图 1.1—图 1.6 为 1336 Impact 变频器内部的部分结构。

图 1.1 IGBTs 电路

输出电流平稳,电动机运行噪声低

图 1.2 主电路板

贴片技术减小了电路板的面积,保持功能不变

为各种应用优化了变频器的 I/O 及其特性

图 1.3　逻辑控制接口板

提供易用的编程、操作和诊断信息

图 1.4　人机接口模块

允许和一切 SCANport 编程终端与通信块连接

图 1.5　网络通信接口

使用插入式语言模块，支持多种语言

图 1.6　语言模块

§1.3　1336 Impact 变频器的功能特点

1. 简易启动过程

1336 Impact 变频器提供了菜单驱动启动程序，提示用户输入电动机的铭牌数据和运行自整定试验程序。同时，通过协助用户组态继电器输出和 L 控制（可选卡），完成数据 I/O 口的设置。还可协助链接 I/O 和基准源，进而实现模拟 I/O 口的设置。

2. 速度控制

1336 Impact 变频器通过使用宽频带的速度调节器，实现精准的速度控制，从而在使用编码器且速度范围为 1000∶1 时，在 100∶1 的速度范围内调速精度为 0.001%，无编码器运行调速精度为 0.5%，范围可达 120∶1。

3. 转矩控制

使用磁场定向技术和电流调节型 PWM（CRPWM）可实现转矩的控制。用户可编程使变频器采用速度控制、转矩控制或两者同时控制，可调的转矩、功率和电流上限给予控制应用极大的灵活性。

4. 过程微调控制器

1336 Impact 变频器内置过程微调控制器，可集成跳动量、负载传感器或其他张力变送设备的功能，无需外部控制器，并可微调变频器的速度和转矩输出，包括：

① 可编程的 PI 调节器。

② 可选的速度或转矩求和运算。

③ 可编程的噪声滤波器。

5. 减少电压反射

1336 Impact 变频器内部装配内置电压反射抑制器，其功能是降低电压过渡过程的时间，从而减少电机的过压时间。

6. 电子式 I2t 电动机保护

电子式电动机过载保护，可在单电动机应用场合省去外部过载保护设备，其特点如下：

① 电动机的保护可用参数编程。

② 电动机的保护是基于电动机的额定数据，而不是变频器的额定数据。

③ 脱口状态的过载报警优先于电动机的关断。

④ 通过 UL 认证，符合 NEC 标准的 430 条。

7. 电动机温度补偿

当电动机温度发生变化时，电动机的输出转矩也会变化。一般来讲，电动机内部需安装温度传感器，温度传感器将电动机温度反馈到变频器中，变频器根据电动机温度进行转矩控制的调节。1336 Impact 变频器无须特别在电动机内安装温度传感器就可自动补偿电动机温度的变化。

8. 功能块编程

功能块增强了变频器使用的灵活性和便捷性，为具体应用编程提供了基本功能。1336 Impact 变频器具有 11 个功能块，任意 1 个功能块均可配合 17 个功能参数使用，如图 1.7 所示。1336 Impact 变频器可用的功能块如下：①时间延迟；②逻辑加/减法；③逻辑乘/除法；④磁滞；⑤加/减运算；⑥乘/除运算；⑦频带；⑧最大/最小值；⑨定标；⑩固定状态机器；⑪加法/减法计数器。

9. 转矩、功率和电流极限

1336 Impact 变频器具有独立的参数极限值，包括控制电动机转矩输出极限值、功率输出极限值和电动机电流极限值，每种极限值均可通过编程的方式独立设定。

§1.4　1336 Impact 变频器的接口与通信

1.4.1　逻辑接口可选卡

逻辑接口可选卡(L-可选件)提供 1336 Impact 变频器的各种方式的信号和控制命令。通过编程可以实现各种输入的组合，以适应具体现场安装的要求。可用的 6 种不同的可选卡如下：

① L4：触点闭合接口。

② L5：+24V AC/DC 接口。

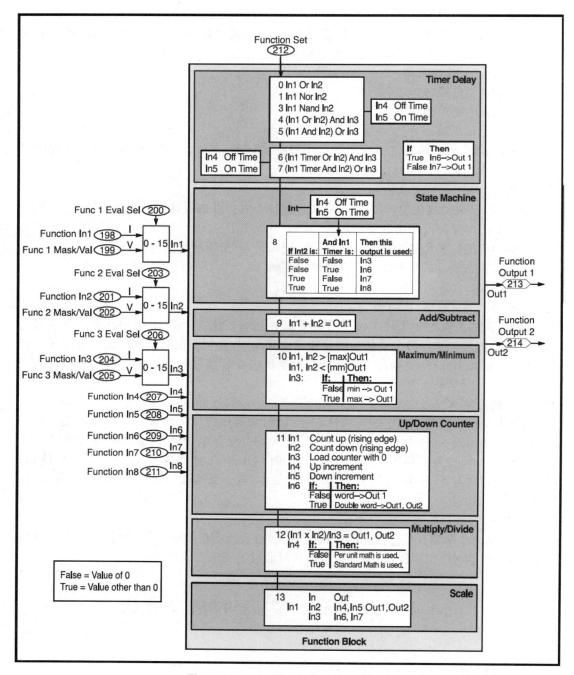

图 1.7 1336 Impact 变频器功能块概况

③ L6：115V AC 接口。
④ L7E：带编码器反馈信号的触点开关量输入接口。
⑤ L8E：+24V AC/DC 带编码器反馈输入信号的接口。
⑥ L9E：115V AC 带编码器反馈输入信号的接口。

本书实验实训篇中 1336 Impact 变频器内部可选卡型号为 L7E。

1.4.2 人机接口模块

1336 Impact 变频器具有功能强大、操作简便的人机接口模块（Human Interface Module，HIM），用户通过人机接口模块进行各种设置、操作和诊断。

HIM 的主要功能和特点如下。

1. 功能

HIM 具有上传和下载功能。用户可以获得和存储变频器的所有参数，还可以将一台变频器的参数传输并存储到其他变频器中。

2. 灵活性

HIM 除了有编程和图形显示功能外，还可以配上控制面板实现变频器的本地控制。标准的配置不仅包括变频器上安装的面板，还有手持式终端等，具体如下：

① 数字式加/减速键：速度控制，启动/停止，点动，方向选择和方向指示。
② 模拟量电位器：速度控制，启动/停止，点动和反向。
③ 编辑器：仅为一个空的控制面板，不能本地控制，仅提供编程和故障检修功能。
④ 安装面板：在 NEMA 标准 1 类和 NEMA 标准 12 型/UL4X-室内型中可选用，其可选型组件能把该模块安装在变频器机盖的门上。

本书实验实训篇中 1336 Impact 变频器人机接口模块如图 1.8 所示，其控制面板为数字式加/减速键。

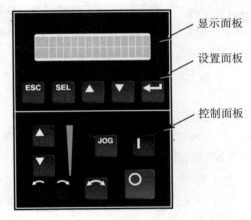

图 1.8　人机接口模块

3. 图形编程终端

图形编程终端（GPT）具有一个全数字键盘，可进行高级参数编程和参数的上传与下载。

1.4.3 网络通信

1. 串口通信

通过可选的 SCANport 通信模块，1336 Impact 变频器可以与其他通信网络相连，如图 1.9 所示。

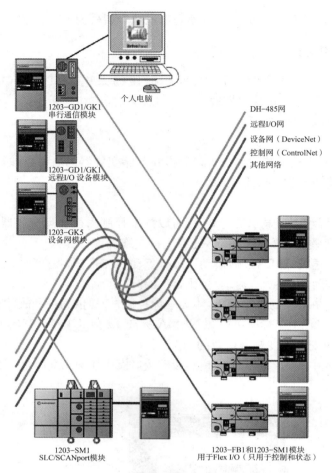

图1.9 1336变频器的网络通信

具体网络如下：

① 设备网(DeviceNet)：设备网连向 SCANport 模块(适用于所有变频器)。
② 控制网(ControlNet)：控制网连向 SCANport 模块(适用于所有变频器)。
③ RS-232/422/485：DF 1 协议与 DH-485 协议(用户自选的协议)。
④ 远程 I/O：远程 I/O 单条分支线连接到 PLC，支持整块传输和链路模式的离散传输。
⑤ Flex I/O：Flex I/O 连向 SCANport 模块(适用于所有变频器)。
⑥ SLC：SLC 连向 SCANport 模块(适用于所有变频器)。
⑦ 外围设备：支持 SCANport 协议，最多可连接 6 个并行设备。

其他 SCANport 产品，如 1305/1336 PLUS/1336 FORCE/1397/1557 变频器，都可以和 SCANport 通信模块一同使用。

2. 支持 Windows 系统的变频器软件

变频器面板(DrivePanel 32)和变频器管理器(DriveManager 32)是变频器工具箱(DriveTools 32)的一部分，是基于微软 Windows 系统的产品，允许对 A-B 公司的交流直流数字化拖动装置进行编程、监控和维护。

3. A-B 公司的远程 I/O 网
① 变频器可设置为可编程控制器(PLC)的一个远程 I/O 逻辑机架。
② 不需任何特别编程,就可以直接使用 PLC I/O 映像表进行实时数据传输。
③ 变频器可组态为 1/4,1/2,3/4 或 1 个逻辑机架。
4. Flex I/O 模块
① 为 1336 Impact 变频器或其他 A-B 公司的 SCANport 产品,提供与 Flex I/O 系统的数字连接。
② 简化安装,缩减连线。
③ 一个 Flex I/O 适配器可以连接 8 个 SCANport 产品。

§1.5　1336 Impact 变频器的参数与设置

1.5.1　变频器的内部参数

1336 Impact 变频器的内部参数被分成 7 个文件,以帮助简化编程和用户访问。与此同时,这些文件被细分成组,每个组中的参数是功能相关联的特定元素。图 1.10 为 1336 Impact 变频器的整体参数目录树,图 1.11—图 1.13 列出了"Program"下每个文件中可以看到的组和具体参数。

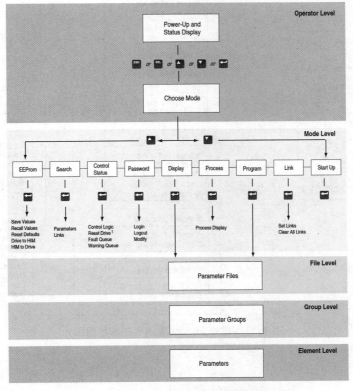

图 1.10　1336 Impact 变频器的整体参数目录树

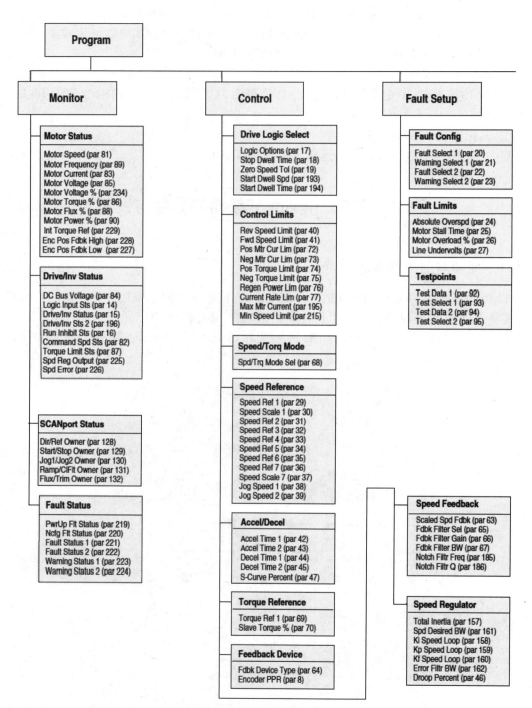

图 1.11　1336 Impact 变频器的参数组(1)

Interface/Comm

Digital Config
- L Option Mode (par 116)
- L Option In Sts (par 117)
- Relay Config 1 (par 114)
- Relay Setpoint 1 (par 115)
- Relay Config 2 (par 187)
- Relay Setpoint 2 (par 188)
- Relay Config 3 (par 189)
- Relay Setpoint 3 (par 190)
- Relay Config 4 (par 191)
- Relay Setpoint 4 (par 192)
- Mop Increment (par 118)
- Mop Value (par 119)
- Pulse In PPR (par 120)
- Pulse In Scale (par 121)
- Pulse In Offset (par 122)
- Pulse In Value (par 123)

Analog Inputs
- An In 1 Value (par 96)
- An In 1 Offset (par 97)
- An In 1 Scale (par 98)
- An In 1 Filter (par 182)
- An In 2 Value (par 99)
- An In 2 Offset (par 100)
- An In 2 Scale (par 101)
- An In 2 Filter (par 183)
- mA In Value (par 102)
- mA In Offset (par 103)
- mA In Scale (par 104)
- mA Input Filter (par 184)

Analog Outputs
- An Out 1 Value (par 105)
- An Out 1 Offset (par 106)
- An Out 1 Scale (par 107)
- An Out 2 Value (par 108)
- An Out 2 Offset (par 109)
- An Out 2 Scale (par 110)
- mA Out Value (par 111)
- mA Out Offset (par 112)
- mA Out Scale (par 113)

SCANport Config
- SP 2 Wire Enable (par 181)
- SP Enable Mask (par 124)
- Dir/Ref Mask (par 125)
- Start/Jog Mask (par 126)
- Clr Flt/Res Mask (par 127)

SCANport Status
- Dir/Ref Owner (par 128)
- Start/Stop Owner (par 129)
- Jog1/Jog2 Owner (par 130)
- Ramp/ClFlt Owner (par 131)
- Flux/Trim Owner (par 132)

SCANport Analog
- SP An In1 Select (par 133)
- SP An In1 Value (par 134)
- SP An In1 Scale (par 135)
- SP An In2 Select (par 136)
- SP An In2 Value (par 137)
- SP An In2 Scale (par 138)
- SP An Output (par 139)

Gateway Data In
- Data In A1 (par 140)
- Data In A2 (par 141)
- Data In B1 (par 142)
- Data In B2 (par 143)
- Data In C1 (par 144)
- Data In C2 (par 145)
- Data In D1 (par 146)
- Data In D2 (par 147)

Gateway Data Out
- Data Out A1 (par 148)
- Data Out A2 (par 149)
- Data Out B1 (par 150)
- Data Out B2 (par 151)
- Data Out C1 (par 152)
- Data Out C2 (par 153)
- Data Out D1 (par 154)
- Data Out D2 (par 155)

Motor/Inverter

Motor Nameplate
- Nameplate HP (par 2)
- Nameplate RPM (par 3)
- Nameplate Amps (par 4)
- Nameplate Volts (par 5)
- Nameplate Hz (par 6)
- Motor Poles (par 7)
- Service Factor (par 9)

Encoder Data
- Encoder PPR (par 8)

Inverter
- PWM Frequency (par 10)
- Inverter Amps (par 11)
- Inverter Volts (par 12)

Motor Constants
- Stator Resistnce (par 166)
- Leak Inductance (par 167)
- Flux Current (par 168)
- Slip Gain (par 169)
- Motor Poles (par 7)

图 1.12　1336 Impact 变频器的参数组(2)

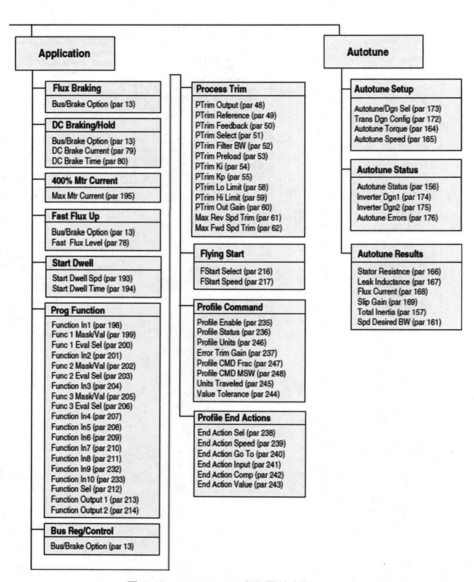

图 1.13　1336 Impact 变频器的参数组(3)

1.5.2 参数描述

参数手册中对每个参数都有具体的描述,其格式如图 1.14 所示。

Par#	Parameter Name		
	Parameter Description		
Parameter Number	1		#
File: group	2		file and group
Parameter type	3		destination or source
Display	4		user units
Factory default	5		drive factory setting
Minimum value	6		minimum value acceptable
Maximum value	7		maximum value acceptable
Conversion	8		drive units = display units
Enums	9		values

图 1.14 参数描述

1—参数编号:每个参数分配一个唯一的编号。编号用于向该参数读写信息。

2—文件:组:列出参数所在的文件和组,一个参数可以列在多个文件和组中,有的参数不能列在任何文件或组中,并且必须通过线性列表访问。

3—参数类型:有 3 种参数类型。

源:该值仅由驱动器更改,用于监视值。

目标:该值可通过编程更改,目标是常量。

可链接目的地:这个值可以是指向另一个参数的链接,也可以是一个常量。

4—显示:用户在显示器上看到的单位,如位、赫兹、秒、伏等。

5—工厂默认值:在工厂中定义的每个参数的默认值。

6—最小值:参数可能的最小设置值。

7—最大值:参数可能的最大设置值。

8—转换:用于与串行端口通信的内部单元,并合理读取或写入驱动的值。

9—枚举:相关联的文本描述。

1.5.3 参数设置方法

利用操作面板,1336 Impact 变频器的内部参数设置通过人机接口模块的显示面板与设置面板共同完成(设置面板的具体功能描述详见实验实训篇)。

在网络通信情况下,1336 Impact 变频器内部参数的观测与设置在上位机的控制标签内实现,在控制标签内可以直接观测当前版本下的所有参数,并可根据具体的实验需求进行更改与设置。

需要注意的是,面板操作与通信操作不可以对参数同时进行设置。

§1.6 1336 Impact 变频器的故障检修

1336 Impact 变频器指示三级故障,并且某些故障可设置报警或切断变频器。易读的故障文字显示在 HIM、GPT 上,或通过 DriverTools 软件显示。变频器同时将故障以 LED 代码序列显示在控制板上，如图 1.15 所示。

图 1.15　故障显示

与此同时，单独的故障缓冲区和报警缓冲区可以储存最近的 32 条信息，并且缓冲区维护在非易失性内存中，具有独立的清除故障和清除缓冲区功能，如图 1.16 所示。

故障缓冲区	
故障码	故障信息
1　02016	过压
2　02018	接地故障
3　03025	绝对速度过大
.	
.	
.	
32	

报警缓冲区	
报警码	报警信息
1　01083	电动机接近过载
2　05086	外部故障输入
3　01085	电动机失速
.	
.	
.	
32	

图 1.16　故障/报警缓冲区

§1.7　1336 系列变频器的应用

从基于 IGBT 的电源结构到先进的微处理器，A－B 公司的 1336 系列变频器采用了先进的电气技术，从而使 1336 变频器能够提高效率和节约能源。

1. 物料传输

1336 Impact 变频器采用 FORCE 技术，适用于某些坚硬物料的传送，比如物料分类、转移、装载等传送带的控制。变负载要求、零速满转矩和精确速度控制，对于 1336 Impact 变频器而言是很容易处理的，而速度/转矩选择、预置速度、降速控制、过程调整和设备链接等功能保证了应用的灵活性。FORCE 技术的转矩控制可防止由传送带的冲击负载引起的过流脱扣。同时，1336 Impact 变频器具有定长剪切应用中要求的宽频带和快速响应性。

2. 挤压机和搅拌机

1336 Impact 变频器最初是为满足挤压机和类似应用的高要求而设计的。某些特殊要求可能是从零到满速恒转矩，启动时转矩大于额定值但无脱扣。1336 Impact 变频器具有过流能力，电动机上限电流可编程至 400% 的额定电流。使用可选的编码器接口，1336 Impact 变频器的调速精度达 0.001%。这就有助于确保某些搅拌和计量应用中要求的精准速度控制。

3. 特殊应用

由于拥有灵活的参数和选项，1336 Impact 变频器适用于大部分的特殊应用中，像风机、机床和快速旋转设备。可编程 I/O、功能块、可选的编码器反馈、速度控制、转矩控制和通信选件，使该变频器可处理大部分的特殊应用。除标准变频器产品外，变频器程序允许 OEM（原始设备制造商）将 1336 Impact 变频器装入其机箱中，以处理特殊的应用，如升降机、试车台。

第 2 章 PowerFlex 系列变频器介绍

§2.1 PowerFlex 系列变频器的功能特点

Allen-Bradley 公司的 PowerFlex 系列变频器提供广泛的控制模式,几乎能满足任何电动机的控制需要。通过功能、选项和包装组合实现应用通用性,满足安全要求、编程以及配置的方便。

2.1.1 可扩展的电动机控制

由于存在各种各样的应用需求,PowerFlex 系列变频器提供广泛的电动机控制解决方案。从开环速度调节到精确速度和转矩控制,PowerFlex 系列变频器可以满足从最简单到最苛刻的应用。该系列还具有多种硬件、软件、安全和包装选择,满足各种实际需求。

① 选择为应用需求而制造的变频器以及与应用需要同样多的选项,从而降低总的使用成本。

② 通过特定应用控制,例如适用于起重应用的 TorqProv 以及适用于油井的关泵功能,提升运行效率。

③ 通过高级诊断以及异常工作参数通报,避免计划外的停工时间。

④ 通过软件工具和向导方便地配置和试车。

2.1.2 无缝的变频器和控制系统集成

通过无缝地集成 PowerFlex 变频器和 Logix 可编程自动化控制器,节省了配置和故障处理的时间。

① 通过 EtherNet/IP、DeviceNet、ControlNet 和其他网络统一了工厂基层和总部之间的通信,并可方便地访问实时信息以及生产数据。

② 通过一个软件工具整合变频器系统配置、运行和维护,降低了编程、安装和整体拥有的成本。

③ 配置数据使用单个储存库,可方便地访问系统和设备级别的数据和诊断信息,从而提升效率。

2.1.3 电动机控制

为获得最佳的电动机控制解决方案，PowerFlex 系列变频器利用了大量的控制技术，让用户几乎能够满足从开环速度调节到精确转矩和速度控制的任何应用需求。除了工业标准电动机控制以外，PowerFlex 系列变频器提供独特的控制技术，为用户带来更大的应用灵活性。

① FORCE 技术。矢量控制与受专利保护的 FORCE 技术相结合，提供绝佳的低速/零速性能，带给用户精确和可靠的转矩调节和速度控制。

② DeviceLogix。DeviceLogix 是部分 Allen-Bradley 产品中采用的嵌入式控制技术，可控制输出以及管理设备上的状态信息。采用 DeviceLogix 技术的变频器通过控制输出以及管理驱动器内的状态和信息，帮助提高系统性能和效率。

③ DriveLogix。带有 DriveLogix 选项的 PowerFlex 700S 交流变频器提供一个嵌入式 Logix 处理器，能在驱动系统或单独应用中为重要控制提供最佳集成。

④ SynchLink。在 PowerFlex 700S 变频器中提供的变频器到变频器数据链路是一种高速、同步、变频器到变频器间的数据链路，用于传输同步的变频器和应用数据。

2.1.4 特定应用控制

部分 PowerFlex 变频器具有专门化的变频器参数，用于支持特殊应用。

① 定位。PowerFlex 40P/700/700S/755 为单轴应用进行了优化，其功能从简单的定位和速度仿形以及点到点规划到复杂的电子齿轮、注册、复位和安全能力。

② 提升。帮助确保任何提升或起重应用中的负载控制。这种高级控制功能确保在任何移动命令过程中机械制动在停止变频器时能够控制负载，释放制动时变频器能够控制负载。

③ 关泵。这种专门用于油井应用的独特功能是一种受专利保护的关泵参数，通过测量电动机的转矩和电流来确定油井流量。这种对传统机械流量计的替代方案可让泵的操作人员根据油井流量来优化产量，同时又能通过保护钻杆和电动机资产来减少停工时间。

2.1.5 安全功能

通过行业领先的安全选项，PowerFlex 系列变频器可提升效率并帮助保护人员。

① 保护避免出现潜在有害的设备或工作状况。

② 通过不需要使用外部继电器的安全速度监控选项，降低了成本以及接线复杂度。

③ 在发生需要安全系统的情况后，能更快地恢复生产。

④ 满足安全额定要求，最高可达并包括 PLe/SIL 3 以及 CAT 3 和 CAT 4 安全性等级。PowerFlex 40P/70/700H/700S AC/755 变频器提供可选的安全关断转矩功能，可提供安全关断控制。

安全关断转矩对于需要拆除电动机的旋转电源而不要关闭变频器的应用来说，是其安全性的理想之选。安全关断转矩功能提供的优势为：在需要安全系统后可快速启动，有助于减少重复启动所造成的磨损。

在需要控制和监视应用速度时，PowerFlex 755 的安全速度监控选项将安全关断转矩功能与集成的安全继电器功能以及安全速度控制技术组合成一个硬件选项,将安全性等级提升到 PLe/SIL 3 和 CAT 4。通过安全速度监控选项,可以安全地监视和控制应用的速度,让操作人员执行维护工作而不用关掉设备。

具有安全关断转矩功能的变频器可使用 MSR 57P 安全继电器进行配置,以实现相同的安全限速能力和安全性等级,如图 2.1 所示。

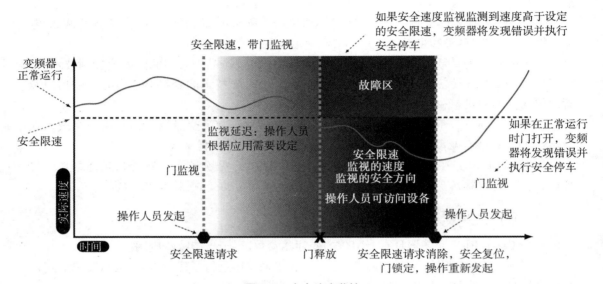

图 2.1 安全速度监控

安全速度监控选项提供以下功能:①安全关断转矩;②停机类别 0,1 和 2;③安全停止;④安全限速;⑤安全最大速度;⑥安全最小加速度;⑦安全方向;⑧零速监视;⑨门控制和监视;⑩启用开关输入。

2.1.6 网络通信

PowerFlex 系列变频器使用 NetLinx 开放式网络架构,提供常用的功能和服务集,适用于在 EtherNet/IP、DeviceNet 和 ControlNet 中使用的通用工业协议(CIP)。在单个网络上具备控制、配置和收集数据的能力,简化工厂通信,并可帮助降低总的拥有成本。轻松管理从车间到总部的信息,无缝地集成用户的整个系统,同时又能控制、配置和收集数据,如图 2.2 所示。

图 2.2 网络通信功能

除了 NetLinx 开放式网络架构以外，PowerFlex 变频器能够支持全世界现有的工业协议。

§2.2　PowerFlex 变频器的类型

PowerFlex 系列变频器和 Kinetix 伺服驱动器可以提供从简单的机械控制、高速定位到中压控制等各种解决方案。这些灵活的产品组合可提供多种解决方案，帮助用户了解产品运营的方方面面，并有效提高生产效率。其变频器产品主要分为低压交流变频器、中压交流变频器和直流变频器 3 种。

PowerFlex 低压交流变频器提供了广泛的特性和功能选项，适合用户的各种应用需求：
① 功率范围为 0.25~3 000 hp/0.2~2 200 kW；适合 100~690 V 的全球电压。
② 支持感应和永磁电机类型。
③ 包含通信和安全功能。
④ 可与 Logix 控制系统无缝集成。

PowerFlex 中压交流变频器可提供高效的电机控制，满足各种苛刻应用的需求：
① 电压范围为 2.3~10 kV，电机电流最高达 720 A。
② 提供空气和液体冷却版本。
③ 支持通过高性能的转矩控制实现简单的速度控制。
④ 包含通信、安全和抗电弧功能。
⑤ 可与 Logix 控制系统无缝集成。

PowerFlex 直流变频器可提供灵活且经济实用的直流控制解决方案：
① 功率范围为 1.5~1 400 hp/1.2~1 044 kW；适合 200~690 V 的全球电压。
② 可根据应用灵活配置。
③ 包括适用于起重应用项目的专用转矩控制。
④ 可与 Logix 控制系统无缝集成。

2.2.1　PowerFlex 低压交流变频器

PowerFlex 低压交流变频器系列提供各种控制模式、功能、选件和包装，以及全局电压和多种额定功率。凭借一致的程序结构和通用的操作员界面，PowerFlex 变频器可轻松实现编程和配置，并可减少设置时间、培训和操作。PowerFlex 紧凑型交流变频器提供了一种经济实用的通用解决方案，适用于单机设备级控制应用项目和简单系统集成。PowerFlex 架构级交流变频器提供了广泛的功能和应用项目特定的参数，是高性能应用项目的理想选择。

1. PowerFlex 755T

PowerFlex 755T 变频器旨在提供谐波消除、回馈、公共母线解决方案，从而降低能源成本、增强灵活性并提高生产力（图 2.3）。这些是首批提供 TotalFORCE 技术的变频器，该技术使用了多项旨在帮助优化用户系统的专利功能。

图 2.3　PowerFlex 755T 变频器

其主要产品如下：①PowerFlex 755TL；②PowerFlex 755TR；③PowerFlex 755TM。

2. 工业型变频器

PowerFlex 工业型交流变频器是经济实用的解决方案，适用于广泛的全球应用（图 2.4）。这种通用变频器的设计特点是易于使用，具备省时优势并且以紧凑的封装优化面板空间和应用项目的通用性。

图 2.4　PowerFlex 工业型交流变频器

其主要产品如下：①PowerFlex 753；②PowerFlex 755；③PowerFlex 755T；④PowerFlex 70；⑤PowerFlex 700S；⑥PowerFlex 700；⑦PowerFlex 700L。

3. 紧凑型变频器

PowerFlex 紧凑型交流变频器提供了一种简单且经济实用的解决方案，适用于单机设备级控制应用项目或简单系统集成（图 2.5）。这种通用驱动器具有易用的设计特点，提供紧凑型封装，可优化面板空间和应用项目通用性。

图 2.5 PowerFlex 紧凑型交流变频器

其主要产品如下：①PowerFlex 523；②PowerFlex 525；③PowerFlex 527；④PowerFlex 4M；⑤PowerFlex 4；⑥PowerFlex 40；⑦PowerFlex 40P；⑧PowerFlex 400。

2.2.2 PowerFlex 中压交流变频器

A-B 公司的 PowerFlex 中压变频器可满足比以往更广范的应用性能需求。这些易于使用的变频器非常适用于可变扭矩应用，例如简单、独立的离心式风机和泵。

1. PowerFlex 6000 中压交流变频器

PowerFlex 6000 中压交流变频器可以在各种应用项目中实现灵活性。它们适用于新的及改装的可变转矩和恒定转矩应用项目。其中一些应用项目包括需要变速电动机控制（2.3～11 kV）的风机、泵和压缩机。这些变频器具有多种配置，具体取决于电动机电压和区域要求，如图 2.6 所示。

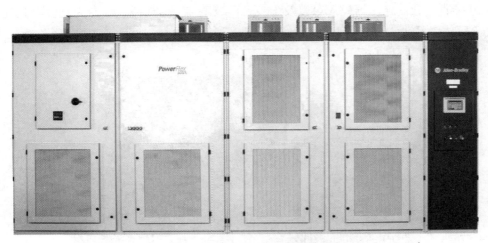

图 2.6 PowerFlex 6000 中压交流变频器

2. PowerFlex 7000 中压交流变频器

PowerFlex 7000 中压交流变频器是风冷或水冷变频器,具有 200~34 000 hp 的宽广功率范围,供电电压范围为 2 400~6 600 V AC。这些独立式通用变频器可控制标准异步或同步交流电机的转速、转矩、方向、启动和停止,如图 2.7 所示。

图 2.7　PowerFlex 7000 中压交流变频器

3. 采用 ArcShield 技术的 PowerFlex 7000 变频器系统

采用 ArcShield 技术的 PowerFlex 7000 变频器系统是首批具有全回馈功能的抗电弧中压变频器。抗电弧系统经过认证,符合最全面的一套全球抗电弧标准。此系统提供高达 50 kA 的电弧故障级别,符合 2B 类可接触性保护标准。

2.2.3　PowerFlex 直流变频器

A - B 公司提供全面的 PowerFlex 直流变频器产品,包括 PowerFlex 直流数字变频器、PowerFlex 直流独立稳压器(SAR)以及 PowerFlex 直流现场控制器,专为直流电机应用设计,改造后功率可达 6 000 hp/4 000 kW,如图 2.8 所示。

图 2.8　PowerFlex 直流变频器

§2.3　PowerFlex 4/40/400 系列变频器

1. PowerFlex 4 交流变频器

PowerFlex 4 交流变频器适合满足全球 OEM 和最终用户对简单性、空间节省和经济实用性的要求。该变频器提供了直观的功能，如带本地电位器和控制键的集成化键盘，这些键开箱即用。

① IP20 NEMA/UL 开放式、板式驱动器，法兰安装和 DIN 导轨。
② 可选 IP30 NEMA/UL 类型 1 机壳转换套件。
③ 最小变频器间距所允许的环境温度高达 50℃(122℉)。
④ Zero Stacking 变频器所允许的环境温度高达 40℃(104℉)。
⑤ V/Hz 控制、滑差补偿。
⑥ 集成式 RS-485 通信。
⑦ 可选串行转换模块，可与采用 DF 1 报文的任何控制器结合使用。
⑧ 可选 ControlNet、DeviceNet、EtherNet/IP、Profibus DP、BACnet 和 LonWorks 通信模块。
⑨ 变频器过载保护、斜坡调节和飞速启动。
⑩ 通过集成式 LCD 键盘或软件进行配置和编程。
⑪ 与 PowerFlex 交流变频器进行的源代码级集成采用 Logix 控制平台来简化参数和标记编程并缩短开发时间。

2. PowerFlex 40 交流变频器

PowerFlex 40 交流变频器在一个易用的紧凑封装中，为 OEM、机器制造商和最终用户提供性能增强的电机控制。采用无传感器矢量控制，以满足低速转矩要求。它们有灵活的防护罩选件，编程简单，可快速安装和配置。

① IP20 NEMA/UL 开放式、板式驱动器，凸缘架，IP66 NEMA/UL 类型 4X/12 和 DIN 导轨。
② 可选 IP30 NEMA/UL 类型 1 转换套件。
③ 最小变频器间距所允许的环境温度高达 50℃(122℉)。
④ Zero Stacking 变频器所允许的环境温度高达 40℃(104℉)。
⑤ V/Hz 和无传感器矢量控制以及过程 PID。
⑥ 集成式 RS-485 通信。
⑦ 可选串行转换模块，可与采用 DF 1 报文的任何控制器结合使用。
⑧ 可选 DeviceNet、ControlNet、EtherNet/IP、Profibus DP、BACnet 和 LonWorks 通信模块。
⑨ 变频器过载保护、斜坡调节和飞速启动。
⑩ StepLogic 可作为独立的位置控制器运行。
⑪ 通过集成 LCD 键盘、远程键盘或软件进行配置和编程。
⑫ 与 PowerFlex 交流变频器进行的源代码级集成采用 Logix 控制平台来简化参数和

标记编程并缩短开发时间。

3. PowerFlex 40P 交流变频器

PowerFlex 40P 交流变频器以紧凑、经济实用的设计,通过 3 类安全转矩中断选件提供闭环控制。该变频器适合满足全球 OEM 和最终用户对灵活性、空间节省和易用性的要求。它可以为转向器、智能传送带、包装机和纺丝机等应用项目实现经济实用的转速控制或基本位置控制解决方案。

① IP20 NEMA/UL 开放式、板式驱动器,法兰安装。
② 可选 IP30 NEMA/UL 类型 1 转换套件。
③ V/Hz 控制、无传感器矢量控制以及过程 PID。
④ 内部 RS-485 和通用工业协议(CIP)通信。
⑤ 4 位数字显示、3 个附加状态指示灯和滚动/复位按钮。
⑥ 可选编程用远程人机接口模块或 PC 接口。
⑦ 带和不带编码器反馈的转速控制。
⑧ 变频器过载保护、斜坡调节和飞速启动。
⑨ 光纤专用特性。
⑩ StepLogic 可作为独立位置控制器运行。
⑪ 通过集成 LCD 键盘、远程键盘或软件进行配置和编程。
⑫ 与 PowerFlex 交流变频器进行的源代码级集成采用 Logix 控制平台来简化参数和标记编程并缩短开发时间。
⑬ 安全转矩中断功能经过认证符合 PLd/SIL 2 CAT 3。

4. PowerFlex 400 交流变频器

PowerFlex 400 交流变频器针对商业和工业风机和水泵的控制进行了优化。在广泛的可变转矩风机和水泵应用项目中,吹扫和阻尼器输入等内置功能可以实现经济实用的速度控制解决方案。可用的一体化 PowerFlex 400 风机和水泵变频器在标准化设计中提供了额外的控制、电源和防护罩选件,是可变转矩风机和水泵应用项目中经济实用的转速控制解决方案。

① NEMA/UL 类型 1,12,3R 和 4 机壳。
② 凸缘架和面板安装选件。
③ V/Hz 控制、滑差补偿和 PID 控制。
④ 内部 RS-485、通用工业协议、嵌入式 Modbus RTU、Metasys N2 和 Apogee FLN P1 车间级网络协议。
⑤ 支持变频器串行端口(DSI)通信模块。
⑥ 变频器过载保护、飞速启动、吹扫和阻尼输入、手动/关闭/自动和休眠/唤醒功能。
⑦ 一体化驱动器选件,包括隔离开关和接触器旁路套件。
⑧ 提供线路电抗器选件。
⑨ 通过集成化键盘、远程键盘或 DriveTools SP 软件进行配置和编程。
⑩ 与 PowerFlex 交流变频器进行的源代码级集成,通过 Logix 控制平台来简化参数和标签编程,并且缩短开发时间。
⑪ 一体化 HVAC 风机和水泵驱动器选件:主输入隔离开关;接触器全旁路,带隔离开

关;接触器基本旁路,带隔离开关。

§2.4 PowerFlex 520 系列变频器

PowerFlex 520 系列变频器集设计新颖和简单实用等特点于一身,可为用户提供一套电机控制解决方案,实现最优的系统性能,缩短设计时间,并交付更高品质的机器。该系列产品包含 3 款变频器,每一款都具备独特的功能集,可满足用户不同的应用需求,其功能特性如图 2.9 所示。

图 2.9 PowerFlex 520 系列变频器功能特性

简化编程:凭借 MainsFree 编程功能,可以通过 USB 连接将配置文件上传和下载到 PowerFlex 525 和 PowerFlex 523 变频器控制模块。

创新设计:得益于此模块化设计,用户可以在安装 PowerFlex 525 或 PowerFlex 523 变频器电源模块的同时配置控制模块。

高工作温度:通过使用控制模块风扇套件并降低电流额定值,可让 PowerFlex 520 系列变频器即使在高达 70℃(158°F)的温度下也能正常运行。

灵活性:PowerFlex 520 系列所有的变频器均可垂直或水平安装,并且两种方向都支持并排安装,水平安装时需要使用控制模块风扇套件。

PowerFlex 523 交流变频器非常适合需要经济高效的电机控制的机器。它的设计目的是协助缩短安装与配置时间,同时提供应用所需的控制。

① 上传/下载变频器配置的标准 USB。

② 通过人机接口模块及 Connected Components Workbench 软件工具简化配置工作。

③ 借助可选的通信模块,可轻松向网络中添加变频器。

PowerFlex 525 交流变频器适合系统集成情况较为简单的机器,并提供包括内置 EtherNet/IP 端口和安全在内的标准功能。

① 可无缝整合于 Logix 控制架构,且支持自动设备配置。

② 灵活的电机控制及安装选项。

③ 借助可选的通信模块,可轻松在网络中添加变频器。

PowerFlex 527 交流变频器旨在与 Allen-Bradley Logix 可编程自动控制器(PAC)配合使用。

① 安全扭矩关断是一项标准功能,可通过硬接线方式实现,也可通过 EtherNet/IP 网络

使用控制器上集成的这一安全功能。

② 内置双端口 EtherNet/IP 支持多种网络拓扑和设备级环网功能。

③ 对于注重简单速度控制和精确电机功能的应用而言，将交流变频器和伺服驱动器配合使用是比较合理的解决方案。

PowerFlex 527 交流变频器可进行简单的速度控制，而 Kinetix 伺服驱动器可处理较为精确的速度、扭矩和位置控制等电机控制操作。

§2.5　PowerFlex 750 系列变频器

PowerFlex 753 和 755 变频器的设计以客户需求为根本，变频器的每一个细节都考虑了用户对灵活性、连接性和生产率的要求，最终得到了一个从最初编程到运行和维护都可为用户带来绝佳使用体验的交流变频器系列。PowerFlex 750 系列变频器比其他同类变频器能提供更多的控制、通信、安全及支持硬件选件，可帮助用户实现生产率最大化。

PowerFlex 753 交流变频器经济实惠、易于使用，适合各种常见的应用。这款变频器标配内置 I/O，还提供 3 个可供通信、安全和附加 I/O 使用的选件插槽。其设计可满足用户应用中最高 250 kW/350 hp 的速度或转矩控制需求。

PowerFlex 755 交流变频器使用简单、具备应用灵活性而且性能卓越。它包含多个控制器、硬件和安全选件。这些变频器具备多种电机控制选项，适用于多种应用。内置的 EtherNet/IP 网络卡可提供实时作业数据，并可轻松整合于 Logix 控制系统中。可理想适用于需要定位、速度或扭矩控制的功率高达 1 500 kW/2 000 hp 的应用。

第 3 章 罗克韦尔自动化网络体系

§3.1 NetLinx 集成架构

罗克韦尔自动化网络是无缝集成于开放网络体系结构的一个部分,称为 NetLinx 开放网络体系结构,它无缝集成了自动化系统的所有组件,从最简单的设备到国际互联网,是基于核心产品和先进通信技术的集成架构。NetLinx 的体系结构如图 3.1 所示,NetLinx 架构下的具体产品如图 3.2 所示。

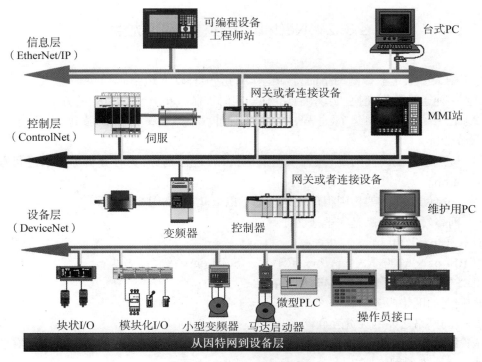

图 3.1 NetLinx 的体系结构

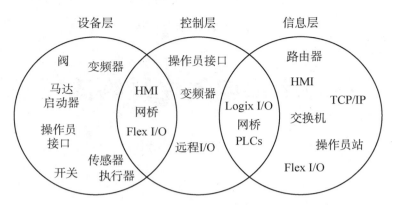

图 3.2　NetLinx 架构下的产品

NetLinx 的体系结构主要包括以下几部分：

（1）Logix 控制平台。多种控制平台上开发的控制执行引擎，共享开发编程环境，提供优选 I/O 连接能力。

（2）NetLinx 通信体系结构。面向实时控制、数据采集和系统组态目标开发的优化的、领先的通信体系结构，在不同的介质类型均能完美实现。

（3）ViewAnyWare 可视化解决方案。从电子操作员终端，到高端工业控制计算机的全系列信息可视化解决方案，共享统一的标准化的开发环境。

§3.2　NetLinx 体系结构的优点

对于绝大多数工业控制网络而言，最基础的应用有以下几点：

（1）一种通过数据交换，实现快速准确的实时控制（Control）的有效的方式。

（2）在系统调试和运行过程中，可以对非关键性系统和设备进行自由的组态（Configuration）。

（3）按照固定的时间间隔，或者根据人机接口、过程趋势、实时分析等要求进行数据采集（Collection）。

NetLinx 网络体系架构则很形象地诠释了以上几点，如图 3.3 所示。

（1）网络对软件产品是完全透明的：

① RSLinx，I/OLinx。

② RSLogix 5000，RSNetWorx，RSView，SoftLogix 等。

（2）网络对硬件产品同样是透明的：

① 相同的网络服务。

② 不同网络之间的无缝互联。

（3）共同的通信服务：控制与通信协议为 NetLinx 开放网络体系结构中的所有三层网络提供一套标准的网络服务。

（4）共同的报文发送服务：允许连接到任何一层网络，另外从任何一层网络实现网络组态和数据采集。

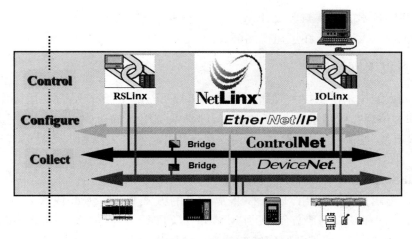

图 3.3 NetLinx 体系的优点

（5）无缝的路由能力：在对整个系统进行组态时，由于可以向下路由或者透过多层网络而无需组态路由表或者在控制器中进行任何额外的编程工作，因此可以节约大量的时间和精力。

（6）共同的知识库：在 NetLinx 体系结构内的不同层的网络具备相似的特点，采用相同的组态工具，可以大大减少所需的培训工作。

（7）基于开放的网络技术：网络全部由开放网络协会组织进行管理，而非罗克韦尔自动化专有。

§3.3 工业以太网理论要点

3.3.1 工业以太网概况

工业以太网（Ethernet）是应用于工业控制领域的以太网技术，在技术上与商用以太网（即 IEEE 802.3 标准）兼容。产品设计时，在材质的选用、产品的强度、适用性以及实时性、可互操作性、可靠性、抗干扰性、本质安全性等方面能满足工业现场的需要。

工业以太网过去被认为是一种"非确定性"的网络，作为信息技术的基础，是为 IT 领域应用而开发的，在工业控制领域只能得到有限应用，这是由于：

（1）工业以太网的介质访问控制层（MAC）协议采用带碰撞检测的载波侦听多址访问/冲突检测（CSMA/CD）方式，当网络负荷较重时，网络的确定性不能满足工业控制的实时性要求。

（2）工业以太网所用的接插件、集线器、交换机和电缆等是为办公室应用而设计的，不符合工业现场恶劣环境的要求。

（3）在工厂环境中，工业以太网抗干扰（EMI）性能较差。若用于危险场合，以太网不具备本质安全性能。

（4）工业以太网不能通过信号线向现场设备供电。

随着互联网技术的发展与普及,Ethernet 传输速率的提高和 Ethernet 交换技术的发展,上述问题在工业以太网中正在迅速得到解决。

3.3.2 工业以太网协议

当工业以太网用于信息技术时,应用层包括 HTTP、FTP、SNMP 等常用协议,但当它用于工业控制时,体现在应用层的是实时通信、用于系统组态的对象以及工程模型的应用协议,目前还没有统一的应用层协议,但受到广泛支持并已经开发出相应产品的有以下几种主要协议。

1. Modbus TCP/IP

该协议由施耐德公司推出,以一种非常简单的方式将 Modbus 帧嵌入 TCP 帧中,使 Modbus 与以太网和 TCP/IP 结合,成为 Modbus TCP/IP。这是一种面向连接的方式,每一个呼叫都要求一个应答,这种呼叫应答的机制与 Modbus 的主/从机制相互配合,使交换式以太网具有很高的确定性,利用 TCP/IP 协议,通过网页的形式可以使用户界面更加友好。

2. PROFINET

针对工业应用需求,德国西门子公司于 2001 年发布了该协议,它是将原有的 PROFIBUS 与互联网技术结合,形成了 PROFINET 网络方案,主要包括:

① 基于组件对象模型(COM)的分布式自动化系统。
② 规定了 PROFINET 现场总线和标准以太网之间的开放、透明通信。
③ 提供了一个独立于制造商,包括设备层和系统层的系统模型。

PROFINET 采用标准 TCP/IP+以太网作为连接介质,采用标准 TCP/IP 协议加上应用层的 RPC/DCOM 来完成节点间的通信和网络寻址。它可以同时挂接传统的 PROFIBUS 系统和新型的智能现场设备。现有的 PROFIBUS 网段可以通过一个代理设备(Proxy)连接到 PROFINET 网络当中,使整个 PROFIBUS 设备和协议能够原封不动地在 PROFINET 中使用。

3. HSE

FF 于 2000 年发布 Ethernet 规范,称为 HSE(High Speed Ethernet,高速以太网)。HSE 是以太网协议 IEEE 802.3、TCP/IP 协议族与 FF H1 的结合体。FF 明确将 HSE 定位于实现控制网络与 Internet 的集成。

HSE 技术的一个核心部分就是链接设备,它是 HSE 体系结构将 H1(31.25 Kbps)设备连接到 100 Mbps 的 HSE 主干网的关键组成部分,同时也具有网桥和网关的功能。网桥功能能够用于连接多个 H1 总线网段,使同 H1 网段上的 H1 设备之间能够进行对等通信而无需 CPU 系统的干涉;网关功能允许将 HSE 网络连接到其他的工厂控制网络和信息网络,HSE 链接设备不需要为 H1 子系统作报文解释,而是将来自 H1 总线网段的报文数据集合起来并且将 H1 地址转化为 IP 地址。

4. EtherNet/IP

EtherNet/IP 是适合工业环境应用的协议体系。它是由 ODVA(Open DeviceNet Vendors Association,开放式设备网络供货商协会)和 ControlNet International 两大工业组织推出的最新成员。与 DeviceNet 和 ControlNet 一样,它们都是基于 CIP(Control and Information Protocol,控制信息协议)的网络。它是一种面向对象的协议,能够保证网络上隐式(控制)的实时 I/O 信息和显式信息(包括用于组态、参数设置、诊断等)的有效传输。

EtherNet/IP 是一个开放式的以太网工业协议,使用标准的以太网 IEEE 802.3、TCP/IP 协议族和 CIP,支持 10/100 Mbit/s 速率。

作为基于 CIP 开发的 EtherNet/IP 产品,除了可以独立应用于控制层和信息层,还可以在设备层得到应用,常常能解决一些特殊场合的需求问题。工业以太网适合用在需要快速响应和大量数据传输的系统,但对环境的要求较之 ControlNet 和 DeviceNet 要高,它不适合放置在较为恶劣的化学环境,以及温度和湿度极端变化、电子噪声强和振动大的场所。

3.3.3 EtherNet/IP 的网络模型

1998 年开始,CI 的一个特别兴趣小组(Special Interest Group,SIG)开始尝试将 DeviceNet 和 ControlNet 所使用的 CIP 移植到以太网上。2000 年 ODVA,CI 和 EIA 3 个国际组织联合推出了 EtherNet/IP。

如图 3.4 所示,EtherNet/IP 是以太网、TCP/IP 与 CIP 的集成,其中应用层使用 CIP 是 EtherNet/IP 和其他工业以太网的主要区别所在。

EtherNet/IP 各层功能如下。

1. 物理层

EtherNet/IP 在物理层和数据链路层采用以太网,其主要由以太网控制器芯片来实现,应用实例如图 3.5 所示。

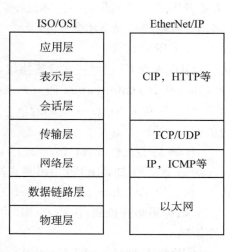

图 3.4 EtherNet/IP 的 ISO/OSI 模型

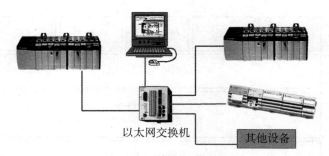

图 3.5 EtherNet/IP 应用实例

2. 数据链路层

和 IEEE 802 标准所规定的其他各种局域网一样,以太网的数据链路层也分为媒体访问控制(MAC)子层和逻辑链路控制(LLC)子层。MAC 子层的主要任务是解决网络上所有的节点共享一个信道所带来的信道争用问题。LLC 子层的任务是把要传输的数据组帧,并且解决差错控制和流量控制的问题,从而在不可靠的物理链路上实现可靠的数据传输。

3. TCP/IP

网络层所要实现的功能是把数据包由源节点送到目的节点。要实现这一功能,网络层需要解决报文格式定义、路由选择、阻塞控制和网际互联等一系列问题。

TCP/IP 模型中的网络层是基于数据报的无连接型的。其工作原理是：源节点的传输层把要传输的数据流分为一个个的数据报，交给网络层；网络层根据一定的算法，为每个数据报单独选择路由；数据报在网络传输的过程中可能会进一步分成多个数据报；每个数据报根据所选定的路由到达目的节点后，由目的节点的网络层拼装成原始的数据报，然后上交目的节点的传输层。

网络层最重要的协议是网际协议(Internet Protocol，IP)。IP 是一个不可靠但会尽力传送的协议。IP 的不可靠体现在它不提供任何核查或追踪功能，因此可能会发生数据报丢失或者出错。IP 的尽力传送体现在它不轻易放弃任何一个数据报，只有在资源耗尽或者网络出现故障的情况下才会放弃。

4. CIP 协议

ODVA 和 CI 已经正式签署协议，共同推动基于 CIP™(通用工业协议)的工业网络，CIP＝Control and Information Protocol(控制与信息协议)。

CIP 是基于 NetLinx 架构的核心网络(DeviceNet，ControlNet，EtherNet/IP)的通用通信协议。

控制协议(Control Protocol)适用于实时 I/O，又称为隐式报文传送(implicit messaging)。例如，RIO 就采用隐式报文传送。

信息协议(Information Protocol)适用于显式报文传送(explicit messaging)，主要用于：组态、数据采集和诊断；在基于 CIP 协议的网络间实现路由功能。例如，DH+就采用显式报文传送。

CIP 为 EtherNet/IP，ControlNet，DeviceNet 网络提供公共的应用层和设备描述，如图 3.6 所示。

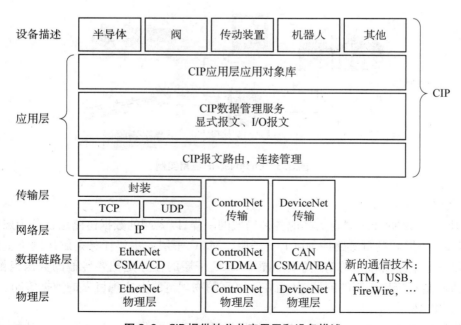

图 3.6　CIP 提供的公共应用层和设备描述

CIP 网络具有如下特点：

① 报文。CIP 根据所传输的数据对传输服务质量要求的不同，把报文分成了两种：显式报文和隐式报文。

② 面向连接。CIP 是一个面向连接的协议，也就是在通信开始之前必须建立起连接，获取唯一的标识符 CID。建立连接时需要用到未连接报文管理器（UCMM）。根据报文的种类不同，连接也分为显式连接和隐式连接。

根据所基于的模型不同，工业网络可以分为两类：基于源/目的地模型和生产者/消费者模型。

在基于源/目的地模型的网络中，每个报文都要指明源和目的地，如图 3.7 所示。发送节点把报文发送到网络中，接收节点根据网络上报文的目的地址段是否与自己的地址相同来判断是否发给自己。该模型的网络只支持点对点通信。

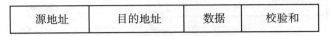

图 3.7　基于源/目的地模型的网络报文格式

在基于生产者/消费者模型的网络中，每个报文都有唯一的报文标识符（MID），格式如图 3.8 所示。在发送报文之前，要在发送节点和接收节点之间建立连接，这样接收节点就知道应该接收的报文的 MID 的具体格式，然后发送节点把报文发送到网络上，接收节点根据报文的 MID 来判断是否发给自己。该模型的网络既支持点对点通信，也支持多播通信。

图 3.8　基于生产者/消费者模型的网络报文格式

3.3.4　EtherNet/IP 技术的应用

EtherNet/IP 网络使用开放式行业标准网络技术提供全厂网络系统，有助于在离散、连续过程、批次、安全、驱动、运动控制以及高可用性应用中实现实时控制并获取实时信息。电机启动器和传感器等设备均可通过 EtherNet/IP 网络连接到控制器、HMI 设备以及企业中。其采用通用网络基础设施，支持非工业和工业通信。

EtherNet/IP 技术是一种可以满足所有工厂策略且使用范围最为广泛的网络，能够处理从最底层到最顶层的通信。其支持以下应用：

① 业务系统/自动化集成：专门采用 IEEE 802.3 和 TCP/IP/UDP 标准。

② HMI：提供大量带宽，可支持几乎所有 HMI 供应商提供的大规模数据密集型 HMI 应用。

③ 设备编程和配置：借助个人计算机上的以太网端口。

④ 对等通信：提供控制器、机器人和其他设备之间的互锁和数据传输功能。

⑤ I/O 控制：更新时间不到 1 ms。

⑥ 时间同步和时间戳：亚微秒级时间协调。

⑦ 驱动控制：配置、监视和协调控制。

⑧ 安全控制：支持标准设备和安全设备在同一网络中运行。
⑨ 运动控制：为最苛刻的协调运动控制应用提供精确的运动控制。

3.3.5 控制系统中的EtherNet/IP模块

EtherNet/IP模块在控制系统中的应用如图3.9所示。

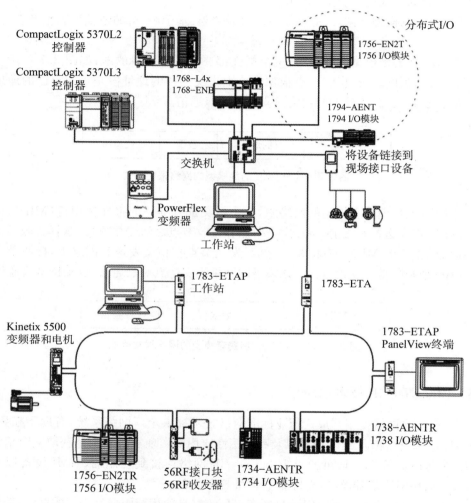

图3.9 EtherNet/IP模块在控制系统中的应用

§3.4 DeviceNet总线理论要点

DeviceNet协议是基于CAN总线技术的，而CAN总线只采用了ISO/OSI网络模型的物理层和数据链路层，DeviceNet在CAN的基础上增加了应用层，扩充了物理层的连接单元接口规范、媒体连接和媒体规范。

DeviceNet 是一种低成本的通信总线。它将工业设备(如限位开关、光电传感器、I/O 设备、马达启动器、过程传感器、变频驱动器、面板显示器和操作员接口等)连接到网络,从而消除了昂贵的硬接线成本。直接互连性改善了设备间的通信,同时又提供了相当重要的设备级诊断功能,这是通过硬接线 I/O 接口很难实现的。DeviceNet 是一种简单的网络解决方案,它在提供多供货商同类部件间的可互换性的同时,减少了硬接线和安装工业自动化设备的成本和时间。

典型 DeviceNet 控制系统结构如图 3.10 所示。

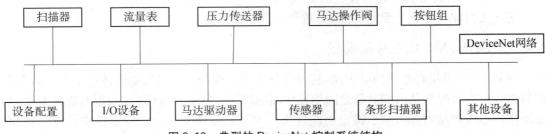

图 3.10　典型的 DeviceNet 控制系统结构

3.4.1　DeviceNet 网络技术规格

DeviceNet 在允许多个复杂设备互连的同时,支持简单设备的互换性。除了可读取离散设备的状态外,DeviceNet 还可以报告马达启动器内的温度、读取负载电流、改变驱动器加减速速率或统计前一小时通过传输带传送的包裹计数。DeviceNet 总线技术规格如图 3.11 所示。

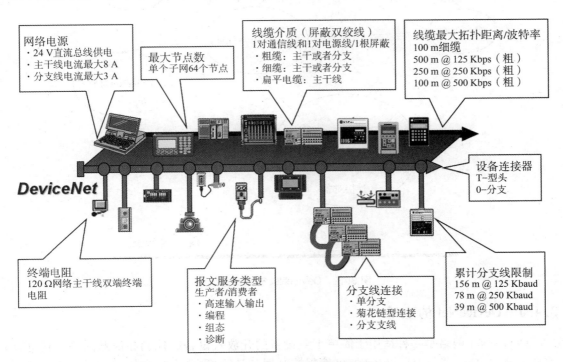

图 3.11　DeviceNet 总线技术规格

3.4.2 DeviceNet 智能接口设计

DeviceNet 智能接口设计的一般步骤：

① 决定为哪种类型的设备设计 DeviceNet 接口，并在 DeviceNet 规范中找到与其对应的设备描述。
② 进行 DeviceNet 接口的硬件设计。
③ 根据 DeviceNet 规范进行软件设计和实现。
④ 决定设备的配置并编写设备的 EDS 文件。
⑤ 完成 DeviceNet 一致性声明，并进行一致性测试。

3.4.3 DeviceNet 节点对象模型

DeviceNet 采用对象建模的方法，将每个总线设备视为一个对象集合体的节点。这些节点的总线行为表现是其内部对象之间相互作用的结果。DeviceNet 协议使用对象的概念，抽象地描述总线产品内部的某个特定功能模块。为了完整地体现一个特定模块具有的特性、功能和运行方式，DeviceNet 协议分别采用属性、服务和行为对一个对象加以描述。图 3.12 描述了 DeviceNet 节点对象之间的关系。

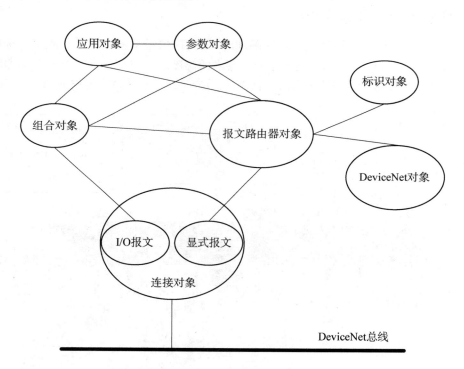

图 3.12 DeviceNet 节点对象模型

3.4.4 DeviceNet 的应用

DeviceNet 网络是一种适用于简单工业设备的开放式设备级控制和信息网络。其可以实现传感器、执行器与上层设备（如可编程控制器和计算机）之间的通信。由于电源和信号共用一根电缆,因此其提供了一种既简单又经济实用的接线选择。

DeviceNet 适合以下类型的应用:①控制低密度的 I/O;②配置设备;③电机控制中心(MCC);④安全控制。

3.4.5 控制系统中的 DeviceNet 模块

DeviceNet 模块在控制系统中的应用如图 3.13 所示。

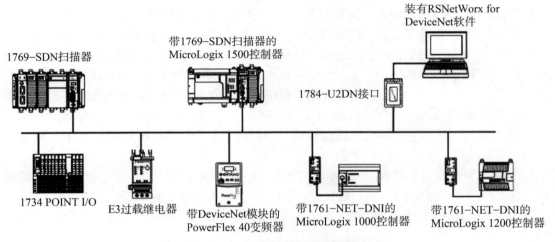

图 3.13 DeviceNet 模块在控制系统中的应用

§3.5 ControlNet 总线理论要点

3.5.1 ControlNet 历史及发展

ControlNet 的技术基础是在 Rockwell Automation 企业长期自动化技术研究过程中发展起来的,最早于 1995 年 10 月面世。1997 年 7 月,Rockwell 等 22 家企业联合发起成立了 ControlNet International 组织,Rockwell 将该项技术转让给了该组织。ControlNet International 组织是一个为广大用户和供货厂商服务的非营利性独立组织,主要负责在全世界范围内推广并发展 ControlNet 技术,提供测试软件及独立的合格性测试,出版发行 ControlNet 技术说明和产品目录以及组织设计和使用 ControlNet 的培训等工作。同时 ControlNet International 组织还为供货商组织了一个特别兴趣小组(Special Interest Groups,SIGs)来统一和规范各厂商的产品,以实现各种产品的标准化。

3.5.2 ControlNet 的技术特点

ControlNet 网络特别适用于对时间有苛刻要求的复杂应用场合的信息传输。它对于同一链路上的 I/O、实时互锁、对等通信报文传送和编程操作,均具有相同的带宽;对于离散和连续过程控制,均具有高度确定性和可重复性;网络支持主从、多主和对等系统结构,对输入数据和对等通信数据实行多信道广播。ControlNet 网络具有高吞吐量、资源共享、组态和编程简单、传输介质为同轴电缆和光纤、支持冗余介质、体系结构灵活和安装费用低等特点。当通信速率为 5 Mbps 时,网络刷新时间最小可达 2 ms,可寻址节点数可达 99 个,同轴电缆传输距离最长可达 6 km(光纤介质使用中继器可达 30 km)。它通过 ControlLogix Gateway 能实现 ControlNet,EtherNet,DeviceNet,RIO,DH+之间的数据交换,是一种先进的高速、实时控制网络和控制系统平台。ControlNet 网络支持主从通信、多主通信、对等通信或这些通信的任意混合形式,对输入数据和对等通信数据实行多信道广播。通信形式可以组态选择,应用更灵活。对等通信或多主通信的采用,可以提高网络的可靠性,改进网络的性能。

ControlNet 网络的一个重要功能是,在传送对时间有苛刻要求的控制信息(如 I/O 状态和互锁的控制信号等)的同时,其他无时间苛求的信息(如程序的上传/下载等)也能传送。而对时间有苛求的信息(预定信息),在每个网络刷新时间内必须传输,这就保障了控制的实时性和可靠性。ControlNet 总线的技术规格如图 3.14 所示,其技术特点可总结如下:

① 在单根电缆线上支持两种信息传输,一种是对时间有严格苛求的信息,另一种是对时间无苛求的信息和程序上传/下载。

② 采取新的通信技术模式,以生产者/消费者通信模式取代了传统的源/目的通信模式。它支持点对点(Peer to Peer)通信,而且允许同一时间向多个设备通信。

③ 可使用同轴电缆,长度可达 6 km,可寻址节点最多达 99 个,两节点间最长距离达 1 km。

④ 安装简单,扩展方便。

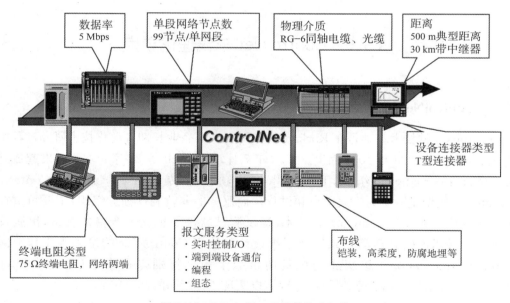

图 3.14 ControlNet 总线的技术规格

3.5.3 ControlNet 技术

ControlNet 现场总线之所以能保障对时间有苛求的信息的有效及时发送、组建高速确定性网络，与它采用独特的通信技术密切相关，下面就从 ControlNet 采用的通信模式、介质访问方式、信息连接方式等方面阐述这一技术。

1. ControlNet 现场总线的通信模式

ControlNet 技术采取新的通信模式，以生产者/消费者(Producer/Consumer)模式取代了传统的源/目的模式。

(1) 显式报文和 I/O 报文。

工厂控制层网络要传送一般的计算机通信网络中需传送的报文，同时需要传送实时的输入/输出(I/O)控制信息及整个控制系统中各控制器互锁信息等。我们用显式报文和 I/O 报文来分别表示。

显式报文用来实现上传和下载程序，修改设备状态，记载数据日志，作趋势分析和诊断等功能，例如我们可以用显式报文对控制器中的 5 个定时器重写预设值或执行自测试操作。它们的结构十分灵活，数据域中带有通信网络所采用的协议信息和要求操作服务的指令。每个节点(设备)必须解释每个显式报文操作所请求的任务，并生成回应。为按通信协议解释这种显式报文，在真正要用到的数据上必须有较大的附加量。这种类型的报文在数据量的大小和使用频率上都是非常不确定的。

I/O 报文在本质上是隐性的(implicit)，因而有时也称为隐式报文，它在数据域中不包括协议信息，仅仅是实时的 I/O 控制数据，这些数据的含义是预定的，因而在节点中处理这些数据所需的时间大大减少。I/O 报文的一个例子是控制器将输出数据发送给一个 I/O 模块(I/O Block, I/O 接口)设备，然后 I/O 模块按照它的输入数据回应给控制器。为了解释这种类型的报文而必须引入的附加量小，数据短，使用频率一致，并且需要高的性能：对 I/O 报文传送的可靠性、送达时间的确定性及可重复性有很高的要求。过去用于 I/O 控制的网络在发送显式报文时不能处理发送数据的时间及报文尺寸上的不确定性因素，控制设备供应商不得不使用不同的网络来满足这两种不同报文类型的不同要求。

(2) 生产者/消费者模式。

ControlNet 技术采取了一种新的通信模式，以生产者/消费者模式取代了传统的源/目的模式。它不仅支持传统的点对点通信，而且允许同时向多个设备传递信息。生产者/消费者模式使用时间片算法保障各节点实现同步，从而提高了带宽利用率。

ControlNet 网络使用生产者/消费者模式交换应用信息。这一模式是理解 ControlNet 通信的基础。在这里，生产者是数据的发送者，消费者是数据的接收者。

生产者/消费者模式是基于开放网络技术的一种新发明的解决方案。该模式允许网络上所有节点同时存取同一个数据，数据一旦产生，便与消费者数量无关，从而使网络系统通信效率更高；生产者/消费者模式还采用多信道广播发送方式，各节点(消费者)可以在同一时间接收到生产者所发送的数据，节点之间接收信息准确同步，网络因此可以连接更多的设备而无须增加网络的通信量。ControlNet 网络支持主从通信、多主通信、对等通信或这些通信的任意混合形式，通信形式可以组态，应用更灵活，对等通信或多主通信的采用可以提高网络的可靠性，改进网络的性能。

(3) I/O 触发机制。

除了传统的轮询方法(Polling)外,生产者/消费者模式还允许用两种新的功能的 I/O 触发方法:状态改变发送(Change-of-State)和周期 I/O 发送(Cyclic)。轮询是从源/目的模型产生的,它本质上是一种两个报文的双向处理(发送方输出数据命令,接收节点收到后做出响应并把反应送回),往往用在主机到它的从机之间,许多轮询周期充满了相同的输入和输出数据,这些冗余的数据浪费了大量的网络带宽。

2. ControlNet 总线的介质访问方式

(1) ControlNet 数据链路层的介质存取控制协议。

ControlNet 采用了一种特殊的令牌传递机制,叫作隐性令牌传递(implicit token passing)。网络上每个节点分配唯一的 MAC 地址(从 1 到 99),像普通令牌总线一样,持有令牌的节点可以发送数据。但是,网络上并没有真正的令牌在传输,相反,每个节点侦听收到每个数据帧的源节点地址,在该数据帧结束之后,每个节点设置一个隐性令牌寄存器(implicit token register),其值为收到的源 MAC 地址加 1。如果隐性令牌寄存器的值等于某个节点自己的 MAC 地址,然后该节点就可以立刻发送数据。因为所有节点的隐性令牌寄存器在任意时刻的值相同,这就避免了冲突的发生。如果某个节点没有要发的数据,则只须发一个空的数据帧(null frame)。

(2) ControlNet 链路层的 NUT 划分。

ControlNet 中被称为网络刷新时间(Network Update Time,NUT)的周期通常分为 3 个主要部分:预定时段、未预定时段和维护时段。

(3) 时间片算法。

时间片算法保证网络上所有节点的同步,如图 3.15 所示。

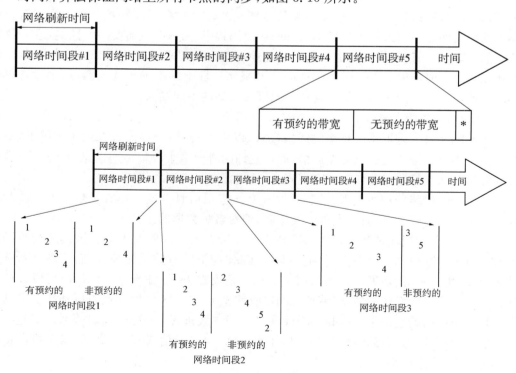

图 3.15　时间片算法

有预约的(scheduled)数据传输：①数据的发送是确定性的,可重复的；②时间关键的I/O以及控制器到控制器之间的互锁(Controller-to-Controller Interlocking)。

非预约的(unscheduled)数据传输：①时间允许就可以发送数据；②非时间关键的点到点报文发送和编程。

有预约的时间段会在既定的每个时间间隔内为组态过的有预约的节点始终保留一次(且仅有一次)发送机会。

非预约的时间段中每个时间间隔的数据发送都会从不同的节点开始。在每个时间间隔内每个节点都会有0次、1次或者多次发送机会,这完全取决于网络负荷状况。

(4) 参数设置。

从上面对MAC层的分析中可以看到,要想使用好ControlNet,其中最主要的是要对NUT,SMAX,UMAX等参数进行设置。用户应根据网络的实际配置情况对它们进行初步计算,然后组态软件。

NUT的大小决定了系统的循环周期：太大,系统的实时性变差；太小,影响了预定时间和非预定时间,使得系统的控制信息和显式报文的发送得不到保证。

由CTDMA控制规则可知：

$$NUT = t_{scheduled} + t_{unscheduled} + t_{maintenance}$$

式中,$t_{scheduled}$,$t_{unscheduled}$,$t_{maintenance}$分别对应系统的预定时间、未预定时间和维护时间。在每一个NUT内,所有scheduled节点都有数据发送,且每个MAC帧中Lpacket部分的长度都达到了其规定的最大值510个字节；同时,在非预定时间段中,CTDMA协议保证了至少要有一个节点发送数据。

$$t_{scheduled} = T \times SMAX$$

$$t_{unscheduled} = T$$

式中,T为节点发送一个最大的MAC帧所需的时间。从上式中可以看到,当网络的结构、长度和所带的节点数等物理特性确定后,$t_{maintenance}$和T值也就确定了,因此NUT的大小主要受SMAX值的影响。由ControlNet的MAC方法可知,网络编址的合理与否对网络性能影响很大,要想很好地使用ControlNet应注意以下几点：

① 对ControlNet进行编址时,应把需要发送实时信息的节点都给予比较低的地址。

② 由于对应于每个空地址,网络都要等待一个槽时间,因此网络上最好不要有比SMAX和UMAX小的空地址。

3. ControlNet的信息连接

在ControlNet上传输的数据可分成非连接(Unconnected)和连接(Connected)两种。非连接信息管理器(Unconnected Message Manager, UCMM)用于在未建立连接的节点间传输信息,这些信息可以是建立连接的请求或简单的非重复性、无时间苛求的数据。所有的节点都要求具有UCMM信息传送和接收能力。UCMM服务用于建立实时数据传送和在节点间传递非连接信息,要求严格的确定性数据则要通过连接信息来传输。

3.5.4 ControlNet的应用

ControlNet网络是一种开放式控制网络,可满足实时、高吞吐量应用的需求。ControlNet支持控制器到控制器的互锁以及I/O、变频器和阀的实时控制。此外,它还可以实现离散和过程应用(包括高可用性应用)中的联网控制。

ControlNet 尤其适合以下类型的应用：
① ControlLogix 平台的网络选项。
② 替换远程 I/O(RIO)网络。
③ 作为多个分布式 DeviceNet 网络的主干网络。
④ 控制器到控制器(对等)的消息传递和互锁。
⑤ 数据采集。
⑥ 高速 I/O 网络。
⑦ 设备配置。

3.5.5 控制系统中的 ControlNet 模块

图 3.16 显示了如何在控制系统中应用 ControlNet 模块。

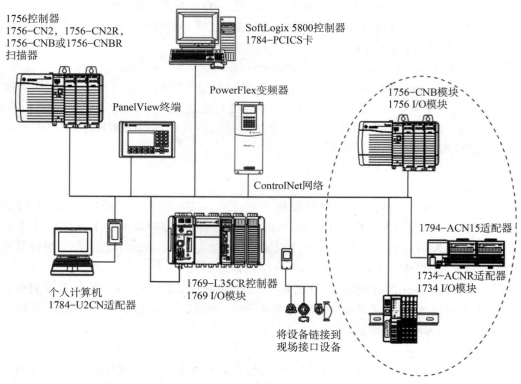

图 3.16 ControlNet 模块在控制系统中的应用

§3.6 不同网络的优势与选型

罗克韦尔自动化所采用的战略是利用开放式网络技术实现从上层管理到下层车间的无缝集成。这些开放式网络使用公共语言,并共享一组通用通信服务。因此,信息可在从下层车间到上层管理平台的整个工厂范围内无缝传送,也可基于 Internet 通信执行电子商务应用,用户可以根据实际需求以及不同网络的各自优势进行随意选择与搭配。

3.6.1 不同网络的特性应用对比

EtherNet/IP,DeviceNet 和 ControlNet 网络支持通用控制和通信服务,可简化设计、启动和维护操作,进而有效降低整个工厂的运营成本。各网络无论是从拓扑结构、容量还是设计特性方面均支持实时控制、设备配置、数据收集和对等联锁功能。这些开放式网络技术受 ODVA 及国际标准组织支持,目前市面上拥有数百种供应商产品,全世界范围内已安装节点达数百万个。表 3-1 为 EtherNet/IP,DeviceNet 和 ControlNet 网络之间的功能特点与应用的对比。

表 3-1 网络功能特点与应用对比

	EtherNet/IP 网络	DeviceNet 网络	ControlNet 网络
功能	与工厂管理系统结合(物料输送);单个高速网络上的组态、数据采集及控制	底层设备直接连接到工厂级控制器,无须通过 I/O 模块转接	支持 PLC 与 I/O 设备之间时间要求严苛的数据传输
典型联网设备	• 计算机主机 • 可编程控制器 • 机器人 • HMI • I/O • 变频器 • 过程仪器 • 无线射频识别	• 传感器 • 电机启动器 • 变频器 • 个人计算机 • 按钮 • 低端 HMI • 条形码阅读器 • PLC 控制器 • 阀组	• 可编程控制器 • I/O 机架 • HMI • 个人计算机 • 变频器 • 机器人
数据重复	大数据包;定期发送数据	小数据包;按需求发送数据	中型数据包;数据传输具有确定性及可重复性
最大节点数	无限制	64	99
数据传输率	10 Mbps,100 Mbps 或 1 Gbps	125 Kbps, 250 Kbps 或 500 Kbps	5 Mbps
典型应用	工厂级架构;高速应用	为底层设备提供电源和连接	冗余应用;确定性通信

3.6.2 EtherNet/IP 网络的优势

1. 极为有效的数据传输

不断增加的波特率(10 Mb,100 Mb),交换机的应用(取代集线器),全双工数据传输,最大限度地减少网络信息的冲突/碰撞,提供与办公用以太网的隔离。

2. 采用现成的(off the shelf)的商用产品和技术

① 同一套安装和支持工具。

② 非常成熟的网络标准,接受程度比较高。
③ 可以充分利用产品的网页浏览(web browsing services)功能。
3. 介质选择
① 主动型总线介质:支持星型拓扑。
② 适应噪声环境(采用光纤)。
③ 采用多种类型的交换机方便地扩展网络长度(铜缆或者光纤)。
④ 密封型介质(IP67),日益涌现的新技术(罗克韦尔自动化作为行业领头羊所开发和推广)。

3.6.3 DeviceNet 网络的优势

1. 较低的采购和安装成本
① 减少现场接线(避免了 I/O 硬接线)。
② 更少的安装、启动和系统维护时间。
2. 网络特性
① 数据流的控制按照生产者/消费者模型进行。
② 连接工厂现场智能设备的能力。
③ 向高层网络提供桥接的能力。
④ 出色的设备级诊断。
⑤ 设备即插即用能力:网络运行中的节点添加/删除。
⑥ 面向少量数据传输进行优化。
⑦ 每个报文(message)8 字节长度。
⑧ 支持对大的数据报进行报文分组(message fragmentation)。
3. 介质选择
① 被动式总线介质,节点加入/退出网络都不会影响系统的运行。
② 密封型(IP67)和非密封型(IP65)介质。
③ 低成本扁平电缆介质。

3.6.4 ControlNet 网络的优势

1. 高速确定性
① 确定性数据传输:知道数据到达的时间,有预约的服务。
② 可重复的数据传输:不论网络上节点设备的多寡,传输的时间是不变的。
2. 冗余
① 可选介质冗余。
② 处理器冗余。
③ PLC-5 热备。
④ ControlLogix 处理器冗余等。
3. 适应工业应用的物理介质
① 适应高噪声环境(同轴电缆或者光纤连接)。
② 本安型介质和产品(防爆型分布式 I/O 模块 Flex Ex,光纤中继器)。
③ 即将推出密封型介质(IP67)。

4. 被动型介质

① 节点加入/退出网络都不会影响系统的运行。

② 支持主干/分支拓扑结构。

3.6.5 通用网络选型指南

用户在选择合适的网络时,一般从以下几个方面考虑:

(1) 应用:确定网络的目的和应用,网络的作用。

(2) 设备:确定需要连接哪些设备及节点类型。

(3) 安装:根据拓扑结构、容量和性能确定哪些网络最符合应用和设备要求。

(4) 易用性:确定哪些网络可以节省开发、调试和维护的时间和成本,确定哪些网络有助于提高生产率。

(5) 成本:网络成本需符合应用要求,确定哪种网络可以提供最大价值。

除此之外,还有若干标准也是选择网络时的参考因素:①应用的复杂性;②拓扑(节点数量、长度、布局);③I/O 点密度;④产品可用性(来自多个厂商、具体设备类型);⑤性能;⑥介质选择(同轴电缆、光纤、网络供电形式);⑦网络组态配置;⑧技术支持、支持能力、支持机构(协会、公司);⑨故障排查工具;⑩MTTR(平均恢复时间);⑪桥接、路由能力;⑫培训(组织、机构)。

EtherNet/IP:更复杂的设备和大的 I/O 模块传送,适用于中等到大规模高性能 I/O 数据和大量报文传送的场合。

DeviceNet:与电气/机械设备安装在一起的简单的现场分布式智能设备和小点数 I/O 模块。

ControlNet:较为复杂的设备,较大点数的 I/O 模块或者机架,适用于高度确定型 I/O 控制,同时又要求一定报文传送的场合。

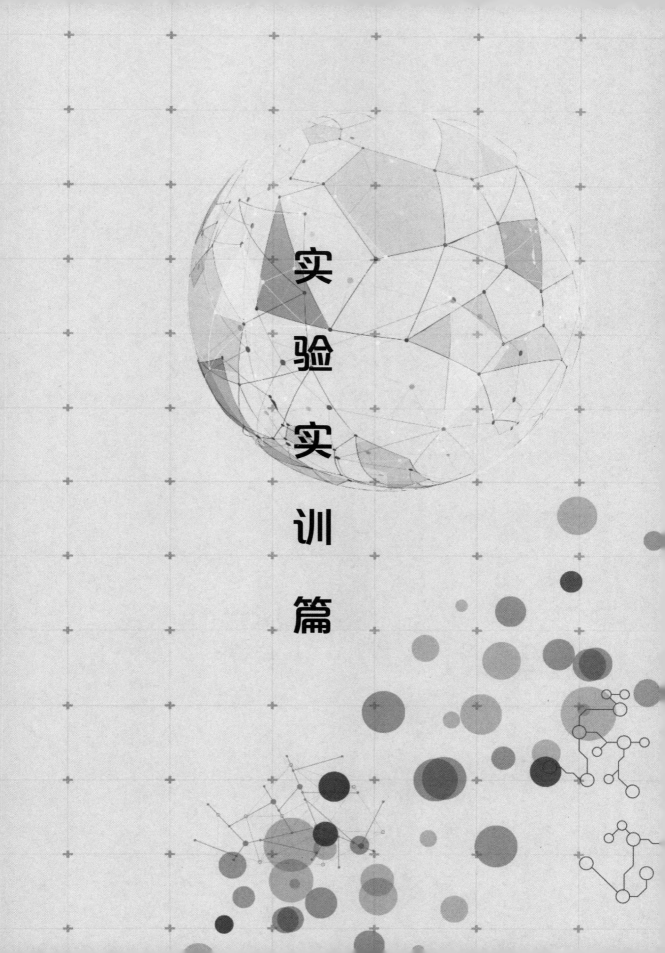

实验实训篇

第4章 1336 Impact 变频器实验

§4.1　1336 Impact 变频器的面板操作控制

实验目的

（1）熟悉变频器的面板操作方法。
（2）熟练掌握变频器的功能参数设置。
（3）熟练掌握变频器的正反转、点动、频率调节方法。

实验内容

（1）熟悉变频器各个输入端子的结构及功能。
（2）了解变频器各个参数的意义，正确设置变频器输出的额定频率、额定电压、额定电流、额定功率、额定转速。
（3）通过操作面板实现电动机启动/停止、正转/反转、加速/减速、点动运行。

实验原理

1336 Impact 变频器人机接口模块如图 1.8 所示。

设置面板键的简要说明如表 4-1 所示。

表 4-1　设置面板键简要说明

图标	功能	名称
ESC	按压此键，程序系统将返回到菜单结构中的上一级	返回键
SEL	按压此键，将使显示屏的顶行或底行轮流变成有效。闪烁的第一个字符表明该行有效	设置键
▲	按压此键，用来增加一个数值或使不同的组（或参数）滚动出现	增键

图标	功能	名称
▼	按压此键,用来减少一个数值或使不同的组(或参数)滚动出现	减键
←	按压此键,可选择一个组或一个参数值,或将一个参数值写入存储器中。在一个参数写入存储器后,显示屏顶行自动变为有效,允许选择另一个参数(或组)	回车键

显示面板键的简要说明如表 4-1 所示。

表 4-1 显示面板键简要说明

图标	描述	名称
I	如果没有其他控制装置发出停止命令,那么此键可使变频器开始运行。"逻辑运算"或"启动屏蔽"可以让此键失去作用	启动键
O	变频器运行时按压此键,会引起变频器按照选择的停止模式停止	停止键
JOG	如果没有其他控制装置发出停止命令,按压此键,变频器将以"点动频率"参数的设定频率值开始点动。释放此键会引起变频器按照选择的停止模式停止	点动键
⌒	按压此键,使电动机以斜坡方式降到 0 Hz,然后以斜坡方式升到反向设置的速度。相应的方向指示灯发光指示电动机的旋转方向	改变方向(正反转)
↶↷	按压此键,可选择一个组或一个参数值,或将一个参数值写入存储器中。在一个参数写入存储器后,显示屏顶行自动变为有效,允许选择另一个参数(或组)	方向 LEDs(指示灯)
▲ ▼	按压此键可增加或减少 HIM 频率命令,可视速度指示器会显示此命令。同时按压上、下箭头,会将频率命令设置成 HIM 存储器中的频率命令值。重新上电或从变频器中拆除 HIM 会将频率命令设置成 HIM 存储器中的频率命令值	上/下箭头(仅对数字速度控制有效)
|	显示速度	速度指示灯(仅对数字速度控制有效)

实验设备与材料

变频器实验设备一套;三相电动机组一组;连接导线若干。

操作方法和步骤

　　启动变频器之前,将 TB3 控制接口板 20～21 端短接
　　(用于启动前变频器内部清错)

1. 恢复工厂默认值

变频器上电,进入操作界面,按参数设置面板任意键,进入模式选择界面,通过▲、▼按键,选择"EEProm",按下↵键,通过▲、▼按键将面板显示调到"Reset Defaults"项,按下↵键,进行复位,保证变频器的内部参数恢复到工厂默认值。

2. 电动机参数设置

为了使电动机与变频器相匹配,需要设置电动机铭牌参数,通过↵与▲、▼按键,将面板显示调到"Program"—"Motor/Inverter"—"Motor Nameplate"中,设置电动机铭牌参数。本实验中电动机铭牌如图 4.1 所示,具体设置见表 4-3。

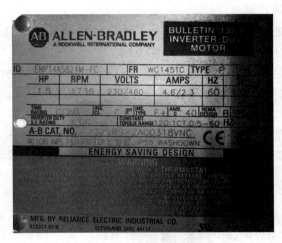

图 4.1 电动机铭牌

表 4-3 电动机铭牌参数设置

参数号-参数名	功能	出厂值	设定值
2 - Nameplate HP	电动机额定功率	30.0 hp	1.5 hp
3 - Nameplate RPM	电动机额定转速	1 750 rpm	1 736 rpm
4 - Nameplate Amps	电动机额定电流	0.2 A	2.3 A
5 - Nameplate Volts	电动机额定电压	460 V	460 V
6 - Nameplate Hz	电动机运行频率	60 Hz	60 Hz
7 - Motor Poles	电动机磁极	4 Poles	4 Poles
9 - Service Factor	运行系数	1.15	1.0

3. 设置变频器内部控制参数

设置面板操作控制参数,通过↵与▲、▼按键,将面板显示调到"Program"—"Control"—"Speed Reference"中,设置参数 38"Jog Speed 1"的值为 400 rpm,见表 4-4。

表 4-4 点动参数设置

参数号-参数名	功能	出厂值	设定值
38 - Jog Speed 1	点动运行速度	100 rpm	400 rpm

4. 变频器运行操作

(1) 变频器启动：在变频器控制面板上按 ▣ 键，启动变频器，变频器将驱动电动机启动。

(2) 加减速运行：电动机的转速（运行频率）可直接通过变频器控制面板上的 ▲、▼ 按键对电动机进行加/减速控制。

(3) 正反转运行：通过按变频器控制面板上的换向键 ▣ 来改变电动机的转向，变频器控制面板上的 ▣、▣ 指示灯将显示电动机的转动方向。

(4) 点动运行：按住变频器控制面板上的点动键 ▣，变频器驱动电动机升速，运行在设置的点动速度值上；松开变频器控制面板上的点动键，变频器驱动电动机降速至零。

变频器的点动运行方向与正反转指示灯方向保持一致。如果按一下变频器控制面板上的换向键，再重复上述的点动操作，电动机在变频器驱动下反向点动运行。

(5) 电动机停止：按变频器控制面板上的停止键 ▣，变频器将驱动电动机降速至零。

记录实验数据

(1) 在变频器参数显示 Mode（参数只读）中观测电动机状态数据并记录。

(2) 通过参数设置面板的 ▣ 与 ▲、▼ 按键，将面板显示调到"Display"—"Monitor"—"Motor Status"中，记录数据。

(3) 通过变频器控制面板的 ▲、▼ 按键，改变电动机转速分别为 300 rpm，800 rpm，1 200 rpm，观察数据变化并记录在表 4-5 中。

表 4-5 数据观测记录

参数号-参数名	功能	电动机转速		
		300 rpm	800 rpm	1 200 rpm
81 - Motor Speed	电动机转速	—	—	—
89 - Motor Frequency	电动机频率	—	—	—
83 - Motor Current	电动机电流	—	—	—
85 - Motor Voltage	电动机电压	—	—	—
86 - Motor Torque%	电动机转矩百分比	—	—	—
88 - Motor Flux%	电动机磁通百分比	—	—	—
90 - Motor Power%	电动机功率百分比	—	—	—

§4.2　1336 Impact 变频器的控制端子应用

实验目的

(1) 掌握变频器基本参数的输入方法。
(2) 掌握变频器输入端子的操作控制方式。
(3) 熟练掌握变频器的运行操作过程。

实验内容

(1) 熟悉变频器各个输入端子结构及功能，以及变频器各个参数的意义。
(2) 通过 HIM 正确设置变频器的控制输入模式。
(3) 根据不同控制模式下 TB3 接口板各端子功能控制变频器对电动机启动/停止、正转/反转、点动等操作。

实验原理

TB3 控制接口板各端子的初始功能定义如图 4.2 所示。TB3 控制接口板端子正常情况下为 19～36 端，其中 19～30 端为变频器常用的控制端子，31～36 端为变频器一些特定的附加功能端子。根据 1336 系列变频器型号以及版本的不同，其 TB3 控制接口板的 31～36 端的功能也各有差别，本实验中 1336 Impact 变频器的型号为 L7E，其 31～36 端为编码器功能的相关接线端。

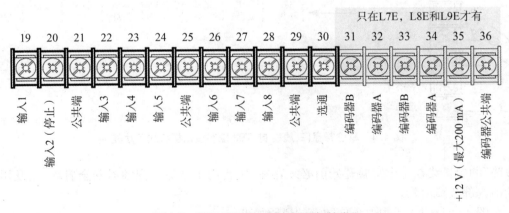

图 4.2　TB3 控制接口板各端子初始定义

TB3 控制接口板的 19～30 端为变频器内部的部分基本功能端，受参数"116 - L Option Mode"输入模式选择的影响，参数 116 的可设置范围为 1～32，设置的值不同，则变频器 19～30 端的端子功能便不同。图 4.3、图 4.4 分别为出厂默认状态与单控制源(三线控制)状态下的 19～30 端子功能定义。

根据出厂状态下 TB3 控制接口板上各端子的功能说明，其中，第 20 端"非停/清除故障"与 30 端"选通端"分别与控制接口板上的公共端(端子 21/25/29)短接，端子 20"非停/清

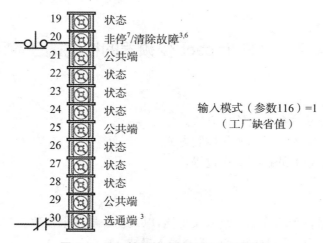

图 4.3 TB3 控制接口板各端子初始功能

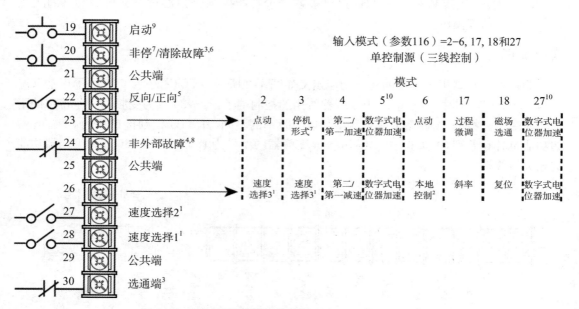

图 4.4 单控制源（三线控制）TB3 控制接口板各端子功能

除故障"的短接线在每个实验开始前必须存在，主要用于被保护变频器安全启动，以及用于启动前内部故障清除。

对图 4.3 与图 4.4 中上角标数字的说明如下。

角标 1：参看速度选择表。

角标 2：变频器要采用本地控制时，必须先停机，并禁止来自其他适配器的控制信号（停机信号除外）。

角标 3：变频器启动前这些输入应存在。

角标 4：在公共母线中，它是预先充电选通的。

角标 5：逻辑选项的第 11 位（参数 17）在反向控制时必须是 0。

角标 6：只适合于软故障。要清除硬故障，需要重复把变频器开机或复位。处理硬故障请参考故障检修的章节。

角标 7：设置停机形式时，参考逻辑选项（参数 17）。

角标 8：该输入在故障清除之前必须存在，变频器将会重新启动。通过故障选择 2（参数 22）和报警选择（参数 23）可将其禁止。

角标 9：设置停机形式时，参考逻辑选项（参数 17）。锁存式（短暂的）启动要求先停止变频器。

角标 10：在方式 5 下，停止后 MOP 值不被置 0；在方式 7 下，停止后 MOP 值被置为 0。

图 4.5、图 4.6 分别为多控制源（三线控制）状态与单控制源（二线控制）状态下的 19～30 端功能定义。其中，单控制源是指只有 TB3 接口板的 23 端和 26 端根据参数 116 的设置值不同而定义不同的功能，而多控制源是在单控制源的基础上将 TB3 控制接口板的 22 端和 27 端也定义新的功能。

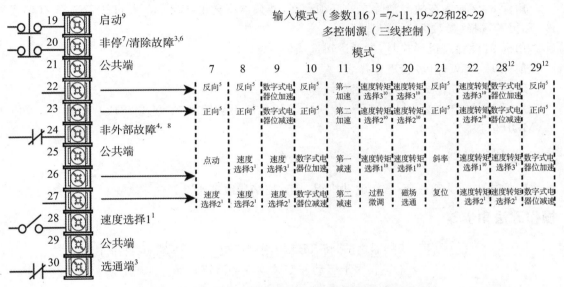

图 4.5　多控制源（三线控制）TB3 控制接口板各端子功能

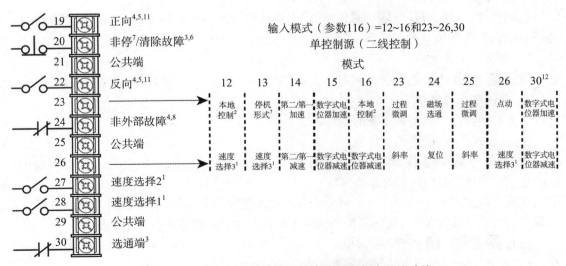

图 4.6　单控制源（二线控制）TB3 控制接口板各端子功能

对图 4.5 与 4.6 中上角标数字的说明如下。

角标 1：参看速度选择表。

角标 2：变频器要采用本地控制时，必须先停机，并禁止来自其他适配器的控制信号（停机信号除外）。

角标 3：变频器启动前这些输入应存在。

角标 4：在公共母线中，它是预先充电选通的。

角标 5：逻辑选项的第 11 位（参数 17）在反向控制时必须是 0。

角标 6：只适合于软故障。要清除硬故障，需要重复把变频器开机或复位。处理硬故障请参考故障检修的章节。

角标 7：设置停机形式时，参考逻辑选项（参数 17）。

角标 8：该输入在故障清除之前必须存在，变频器将会重新启动。通过故障选择 2（参数 22）和报警选择（参数 23）可将其禁止。

角标 9：设置停机形式时，参考逻辑选项（参数 17）。

角标 10：参见速度/转矩选择表。

角标 11：非锁存式（保持型）启动。

角标 12：在方式 9，10 和 15 下，停机后，MOP 值不被置 0；在方式 28，29 和 30 下，停机后，MOP 值被复位为 0。

实验设备与材料

变频器实验设备一套；三相电动机组一组；连接导线若干。

操作方法和步骤

□□启动变频器之前，将 TB3 控制接口板 20～21 端短接□□
□□（用于启动前变频器内部清错）□□

1. 恢复工厂默认值

启动变频器，通过 ■ 与 ■、■ 按键，将面板显示调到"Reset Defaults"项，按下 ■ 键，将变频器的内部参数恢复到工厂默认值。

2. 电动机参数设置

通过 ■ 与 ■、■ 按键，将面板显示调到"Motor Nameplate"项，设置电动机参数。

3. 变频器运行操作

（1）变频器在控制模式 2 下的运行。

① 在模式 2 下，设置相关参数。

通过 ■ 与 ■、■ 按键，将面板显示调到"Program"—"Interface/Comm"—"Digital Config"中，设置控制模式选择参数"116 - L Option Mode"为 2。

注意：当参数"116 - L Option Mode"的值写入变频器后，需要对变频器进行断电操作，重新上电并启动变频器后，其内部控制模式才更新为断电前设置的控制模式值。

通过 ■ 与 ■、■ 按键，将面板显示调到"Program"—"Control"—"Speed Reference"中，设置参数"34 - Speed Ref 5"速度参考 5 为 700 rpm。

通过 ⏎ 与 ▲、▼ 按键，将面板显示调到"Program"—"Control"—"Speed Reference"中，设置参数"38 - Jog Speed 1"点动速度为 400 rpm。

② 按上述设置变频器内部控制参数的方法，设置外部控制模式 2 的参数，见参数设置表 4-6。

表 4-6 控制模式 2 下相关参数设置

参数号-参数名	功能	出厂值	设定值
116 - L Option Mode	控制模式	1	2
38 - Jog Speed 1	点动运行速度	100 rpm	400 rpm
34 - Speed Ref 5	速度参考 5	0 rpm	700 rpm

控制模式 2（参数 116=2），TB3 控制接口板各端子功能如图 4.7 所示。

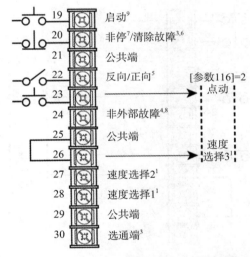

图 4.7 输入模式 2 下 TB3 控制接口板各端子功能

涉及速度选择调节时，请参考变频器内部速度选择表。变频器启动前控制端 20~21 必须提前短接，实验过程中为非停状态。若实验过程中断开 20~21 短接线，则电动机停车。

端子 19 控制电动机启动后，端子 26 短接状态，速度选择 3 闭合，当前转速为速度参考 5 的设置值，端子 20 为非停状态，端子 22 控制电动机正/反转，端子 23 可执行电动机点动状态。

（2）变频器在控制模式 7 下的运行。

① 在模式 7 下，设置相关参数。

通过 ⏎ 与 ▲、▼ 按键，将面板显示调到"Program"—"Interface/Comm"—"Digital Config"中，设置控制模式选择参数"116 - L Option Mode"为 7。

注意：当参数"116 - L Option Mode"的值写入变频器后，需要对变频器进行断电操作，重新上电并启动变频器后，其内部控制模式才更新为断电前设置的控制模式值。

通过 ⏎ 与 ▲、▼ 按键，将面板显示调到"Program"—"Control"—"Speed Reference"中，设置参数"32 - Speed Ref 3"速度参考 3 为 700 rpm。

通过 ← 与 ↑、↓ 按键,将面板显示调到"Program"—"Control"—"Speed Reference"中,设置参数"38 - Jog Speed 1"点动速度为 400 rpm。

② 按上述设置变频器内部控制参数的方法,设置外部控制模式 7 的参数,见参数设置表 4-7。

表 4-7 控制模式 7 下相关参数设置

参数号-参数名	功能	出厂值	设定值
116 - L Option Mode	控制模式	1	7
38 - Jog Speed 1	点动运行速度	100 rpm	400 rpm
32 - Speed Ref 3	速度参考 3	0 rpm	700 rpm

控制模式 7(参数 116=7),TB3 控制接口板各端子功能如图 4.8 所示。

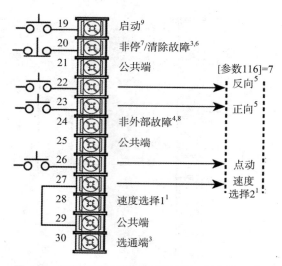

图 4.8 控制模式 7 下 TB3 控制接口板各端子功能

涉及速度选择调节时,请参考变频器内部速度选择表,变频器启动前控制端子 20～21 必须提前短接,实验过程中为非停状态。若实验过程中断开 20～21 短接线,则电动机停车。

端子 19 控制电动机启动后,端子 27 短接状态,速度选择 2 闭合,当前转速为速度参考 3 的设置值,端子 20 为非停状态,端子 22 控制电动机反转,端子 23 控制电动机正转,端子 26 可执行电动机点动状态。

§4.3 1336 Impact 变频器的模拟信号操作与运行（模拟量的单极性控制）

实验目的

（1）掌握变频器基本参数的输入方法。

(2) 掌握变频器的模拟信号控制方式。
(3) 熟练掌握变频器的运行操作过程。

实验内容

(1) 熟悉变频器各个输入端子的结构及功能,以及变频器各个参数的意义。
(2) 掌握模拟量单双极性控制区别,设置模拟量控制参数。
(3) 正确连接模拟控制信号输入,改变模拟信号输入,调节电动机转速大小。

实验原理

1336 Impact 变频器内部提供的模拟量输入/输出以及继电器输出接线端子如图 4.9 所示,对 TB4,TB7 和 TB10 各个端子的说明见表 4-8。

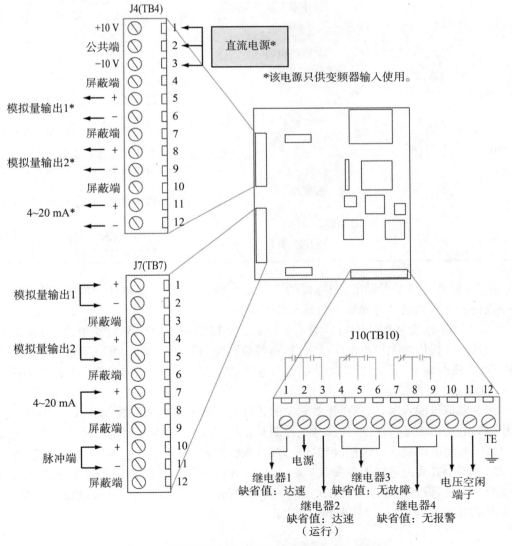

图 4.9 模拟量输入/输出以及继电器输出接线端子

表4-8 TB4,TB7 和 TB10 各个端子说明

端子排	端子号	功能	信号
TB4	4,7,10	屏蔽地	
	1,2,3	直流电源供给	+/−10 V DC 50 mA/V
	5,6,8,9	0～+/−10 V 直流输入	输出阻抗=100 Ω,最大 10 mA
	11,12	4～20 mA 直流输出	输出阻抗=20 Ω
TB7	3,6,9,12	屏蔽地	
	1,2,4,5	0～+/−10 V 直流输入	输入阻抗=20 kΩ
	7,8	4～20 mA 输入	输入阻抗=130 Ω
	10,11	脉冲输入频率基准	+5 V DC 跳线 J8 设置 1-2,+12 V DC 跳线 J8 设置 2-3,必须设置比例系数(脉冲 PPR)不小于 10 mA
TB10	12	逻辑接地,屏蔽	
	1,2,3	可编程触点	
	4,5,6	故障触点	阻性负载=115 V AC/30 V DC,5.0 A 感性负载=115 V AC/30 V DC,2.0 A
	7,8,9	报警端子	阻性负载=115 V AC/30 V DC,5.0 A 感性负载=115 V AC/30 V DC,2.0 A
	10,11	电压空端子	在端子排的逻辑地和其他信号之间提供物理空间

注意:模拟 I/O 是差动的、非隔离的 I/O 点。特别地,把设备的 4～20 mA 输出连接到另外设备的非隔离输入将导致变频器元件的损坏。

1336 Impact 变频器 TB4(J4)端子 1 和端子 2 提供一个±10 V 直流电源。TB7(J7)端子 1 和端子 2 提供两个模拟电压输入口,通过设置变频器控制参数,可应用电位器的调节端改变 TB7 输入端子 1 和端子 2 的模拟输入电压,实现电动机的转速调节。电位器的接线如图 4.10 所示。

在有模拟量输入时,电动机正常转动的正转与反转通常有两种控制方式:一种是由模拟信号控制电动机的运转速度,外部开关控制其正转与反转(单极性控制);另一种方式是通过模拟信号控制电动机的运转速度和运转方向(双极性控制)。

两种控制方式由逻辑选项参数"17-Logic Options"的 11 位决定:第 11 位为 0,变频器控制为单极性,设置速度参考值在 0～1736 rpm 变化;第 11 位为 1,变频器控制为双极性,速度参考值可在−1736～+1736 rpm 变化。

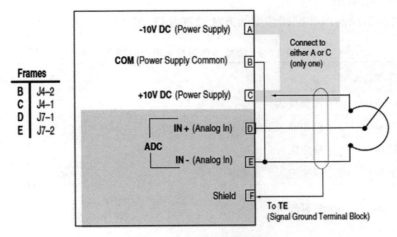

图 4.10 电位器接线说明

实验设备与材料

变频器实验设备一套；三相电动机组一组；连接导线若干。

操作方法和步骤

□□启动变频器之前，将 TB3 控制接口板 20～21 端短接□□
□□（用于启动前变频器内部清错）□□

1. 恢复工厂默认值

启动变频器，通过🔲与🔺、🔻按键，将面板显示调到"Reset Defaults"项，按下🔲键，将变频器的内部参数恢复到工厂默认值。

2. 电动机参数设置

通过🔲与🔺、🔻按键，调到"Motor Nameplate"项，设置电动机参数。电动机参数设置见附录 B。

3. 设置变频器内部控制参数

通过🔲与🔺、🔻按键，将面板显示调到"Program"—"Interface/Comm"—"Digital Config"中，设置控制模式选择参数"116 - L Option Mode"为 2。

注意：当参数"116 - L Option Mode"的值写入变频器后，需要对变频器进行断电操作，重新上电并启动变频器后，其内部控制模式才更新为断电前设置的控制模式值。

通过🔲与🔺、🔻按键，将面板显示调到"Program"—"Interface/Comm"—"Analog Inputs"中，设置参数"97 - An In 1 Offset"为 0，参数"98 - An In 1 Scale"为 2，参数"182 - An In 1 Filler"为 0。

通过🔲与🔺、🔻按键，将面板显示调到"Program"—"Control"—"Drive Logic Select"中，设置参数"17 - Logic Options"的第 11 位为 0。

通过🔲与🔺、🔻按键，将面板显示调到"Program"—"Control"—"Feedback Device"中，设置反馈类型参数"64 - Fdbk Device Type"为无编码器反馈"Encoderless"。

4. 变频器运行操作

按上述设置变频器内部控制参数的方法设置参数(表4-9)。

表4-9 模拟量单极性参考参数设置

参数号-参数名	功能	出厂值	设定值
116 - L Option Mode	控制模式	1	2
97 - An In 1 Offset	模拟输入1偏移量	0	0
98 - An In 1 Scale	模拟输入1比例	2	2
182 - An In 1 Filler	模拟输入1滤波器	0	0
17 - Logic Options	逻辑选项	0001 0000 0000 1000	0001 0000 0000 1000
64 - Fdbk Device Type	反馈方式	Encoderless	Encoderless

设置控制模式2(参数116=2)，TB3控制接口板各端子功能如图4.11所示。

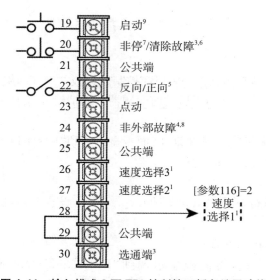

图4.11 输入模式2下TB3控制接口板各端子功能

涉及速度选择调节时，请参考变频器内部速度选择表，变频器启动前控制端20～21必须提前短接，实验过程中为非停状态。若实验过程中断开20～21短接线，则电动机停车。

端子19控制电动机启动后，端子28短接状态，速度选择1闭合，当前转速为速度参考2的值，即模拟量输入电压转换的转速值，端子20为非停状态，调节外接电位器，模拟电压信号在0～10 V变化(顺时针增大，逆时针减小)，对应变频器输出频率在0～60 Hz变化，调节电动机转速在0～1736 rmp变化，端子22控制电动机正/反转。

§4.4 1336 Impact 变频器的模拟信号操作与运行
（模拟量的双极性控制）

实验目的

(1) 掌握变频器基本参数的输入方法。
(2) 掌握变频器的模拟信号控制，以及编码器模式下闭环运行模式。
(3) 熟练掌握变频器的运行操作过程。

实验内容

(1) 熟悉变频器各个输入端子的结构及功能，以及变频器各个参数的意义。
(2) 掌握模拟量单双极性控制的区别，设置模拟量控制参数。
(3) 了解编码器工作原理，掌握编码器接线方法。
(4) 正确连接模拟控制信号输入，改变模拟信号输入，调节电动机转速大小。

实验原理

模拟量输入的单极性与双极性参考由逻辑选项参数"17 - Logic Options"的第 11 位决定：当第 11 位为 0 时，变频器控制为单极性，通过电位器可调的转速值只在 0～1 736 rpm 变化；当第 11 位为 1 时，变频器控制为双极性，通过电位器可调的转速值在 −1 736～+1 736 rpm 变化。

在模拟量单极性参考的控制中，通过电位器分压输入的模拟量范围为 0～10 V，在变频器中通过参数"97 - An In 1 Offset"、参数"98 - An In 1 Scale"和参数"182 - An In 1 Filler"对模拟电压值进行转换处理，最后得到对应的电动机转速范围为 0～1 736 rpm；在模拟量双极性参考的控制中，同样地，通过电位器分压输入的模拟量范围为 0～10 V，在变频器中通过参数"97 - An In 1 Offset"、参数"98 - An In 1 Scale"和参数"182 - An In 1 Filler"对模拟电压值进行转换处理，最后得到对应的电动机转速范围为 −1 736～+1 736 rpm。

通过电位器控制电动机在转速 −1 736～+1 736 rpm 变化时，电动机在经过 0 rpm 及其临界范围时可能会无法正常加/减速，从而停车并震荡。因此，在模拟量双极性参考的实验中，加入编码器反馈，形成速度闭环，这样当电动机转速在 0 rpm 及其临界范围时可以正常停车，并根据实验操作进行加/减速。编码器接线如图 4.12 所示。

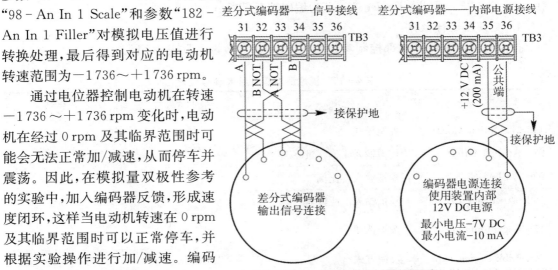

图 4.12 编码器接线

实验设备与材料

变频器实验设备一套;三相电动机组一组;连接导线若干。

操作方法和步骤

□□启动变频器之前,将 TB3 控制接口板 20~21 端短接□□
□□(用于启动前变频器内部清错)□□

1. 恢复工厂默认值

启动变频器,通过 ⏎ 与 ▲、▼ 按键,将面板显示调到"Reset Defaults"项,按下 ⏎ 键,将变频器的内部参数恢复到工厂默认值。

2. 电动机参数设置

通过 ⏎ 与 ▲、▼ 按键,调到"Motor Nameplate"项,设置电动机参数,电动机参数设置见附录 B。

3. 设置变频器内部控制参数

通过 ⏎ 与 ▲、▼ 按键,将面板显示调到"Program"—"Interface/Comm"—"Digital Config"中,设置控制模式选择参数"116 - L Option Mode"为 2。

注意:当参数"116 - L Option Mode"的值写入变频器后,需要对变频器进行断电操作,重新上电并启动变频器后,其内部控制模式才更新为断电前设置的控制模式值。

通过 ⏎ 与 ▲、▼ 按键,将面板显示调到"Program"—"Interface/Comm"—"Analog Inputs"中,设置参数"97 - An In 1 Offset"为 −5,参数"98 - An In 1 Scale"为 4,参数"182 - An In 1 Filler"为 0。

通过 ⏎ 与 ▲、▼ 按键,将面板显示调到"Program"—"Control"—"Drive Logic Select"中,设置参数"17 - Logic Options"的第 11 位为 1。

通过 ⏎ 与 ▲、▼ 按键,将面板显示调到"Program"—"Control"—"Feedback Device"中,设置参数"64 - Fdbk Device Type"为编码器反馈"Encoder"。

4. 变频器运行操作

按上述设置变频器内部控制参数的方法,相关参数设置见表 4-10。

表 4-10 模拟量双极性参考参数设置

参数号-参数名	功能	出厂值	设定值
116 - L Option Mode	控制模式	1	2
97 - An In 1 Offset	模拟输入 1 偏移量	0	−5
98 - An In 1 Scale	模拟输入 1 比例	2	4
182 - An In 1 Filler	模拟输入 1 滤波器	0	0
17 - Logic Options	逻辑选项	0001 0000 0000 1000	0001 1000 0000 1000
64 - Fdbk Device Type	反馈方式	Encoderless	Encoder

设置控制模式 2(参数 116=2)，TB3 控制接口板各端子功能如图 4.13 所示。

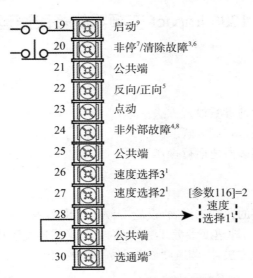

图 4.13　输入模式 2 下 TB3 控制接口板各端子功能

涉及速度选择调节时，请参考变频器内部速度选择表，变频器启动前控制端 20～21 必须提前短接，实验过程中为非停状态。若实验过程中断开 20～21 短接线，则电动机停车。

端子 19 控制电动机启动后，端子 28 短接状态，速度选择 1 闭合，当前转速为速度参考 2 的值，即模拟量输入电压转换的转速值，端子 20 为非停状态。调节外接电位器，模拟电压信号在 0～10 V 变化（顺时针增大，逆时针减小），对应变频器输出频率在－60～60 Hz 变化，调节电动机转速在－1 750～1 750 r/min 变化。

记录实验数据

(1) 在变频器参数显示 Mode(参数只读)中观测电动机状态数据并记录。

(2) 通过参数设置面板的 ↵ 与 ▲、▼ 按键，将面板显示调到"Display"—"Monitor"—"Motor Status"中，记录数据。

(3) 改变模拟量输入，观察数据变化并记录(表 4－11)。

表 4－11　电动机状态数据观测

参数号-参数名	功能	外部模拟量输入(电位器)			
		2.5 V	3.5 V	7.5 V	8.5 V
81 - Motor Speed	电动机转速	—	—	—	—
89 - Motor Frequency	电动机功率	—	—	—	—
83 - Motor Current	电动机电流	—	—	—	—
85 - Motor Voltage	电动机电压	—	—	—	—
86 - Motor Torque%	电动机转矩百分比	—	—	—	—
88 - Motor Flux%	电动机磁通百分比	—	—	—	—
90 - Motor Power%	电动机功率百分比	—	—	—	—

§4.5 1336 Impact 变频器的多段速运行操作

实验目的

(1) 掌握变频器多段速参数设置方法。
(2) 掌握变频器多段速控制方式。
(3) 熟练掌握变频器多段速运行操作过程。

实验内容

(1) 分析变频器内部的速度选择表,设置变频器速度参考 3～7,接入模拟量信号,使其作为其中的一个参考速度,即速度参考 2,接入编码器,形成速度闭环。
(2) 设计多段速运行方法,协调速度参考 3～7 的设置值。

实验原理

当变频器恢复出厂默认状态时,其内部存在的链接也恢复出厂状态,具体如图 4.14 所示。

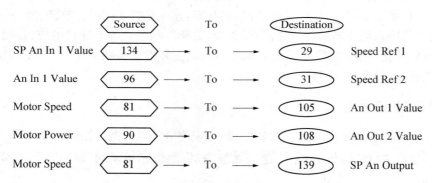

图 4.14 变频器内部初始链接

需要注意的是,参数"134 - SP An In 1 Value"(扫描端子 1 的值)链接到参数"29 - Speed Ref 1"(速度参考 1),即通过变频器 HIM 控制面板上的▲、▼按键实现的电机调速实际为变频器内部参数"29 - Speed Ref 1"的值;参数"96 - An In 1 Value"(模拟量输入 1 的值)链接到参数"31 - Speed Ref 2"(速度参考 2),即改变模拟量输入值实现的电机调速实际为变频器内部参数"31 - Speed Ref 2"的值。

1336 Impact 变频器内部共有 7 个参考速度,分别是参数"29 - Speed Ref 1"、参数"31 - Speed Ref 2"、参数"32 - Speed Ref 3"、参数"33 - Speed Ref 4"、参数"34 - Speed Ref 5"、参数"35 - Speed Ref 6"、参数"36 - Speed Ref 7"。其中,参数 29"速度参考 1"和参数 31"速度参考 2"为初始链接存在的指定的外部给定值,参数 32～36 对应的速度参考 3～7 为内部可设置值,通过 HIM 或网络通信在变频器内部直接写入。变频器内部参考速度的使能状态详见速度选择表 4-12。

表 4-12　变频器内部速度选择表

TB3 控制接口板	端子 26 速度选择 3	端子 27 速度选择 2	端子 28 速度选择 1	使能状态
状态 1	断	断	断	速度参考 1
状态 2	断	断	通	速度参考 2
状态 3	断	通	断	速度参考 3
状态 4	断	通	通	速度参考 4
状态 5	通	断	断	速度参考 5
状态 6	通	断	通	速度参考 6
状态 7	通	通	断	速度参考 7
状态 8	通	通	通	上个状态

实验设备与材料

变频器实验设备一套；三相电动机组一组；连接导线若干。

操作方法和步骤

□□启动变频器之前，将 TB3 控制接口板 20～21 端短接□□
□□（用于启动前变频器内部清错）□□

1. 恢复工厂默认值

启动变频器，通过 ⏎ 与 ▲、▼ 按键，将面板显示调到"Reset Defaults"项，按下 ⏎ 键，将变频器的内部参数恢复到工厂默认值。

2. 电动机参数设置

通过 ⏎ 与 ▲、▼ 按键，调到"Motor Nameplate"项，设置电动机铭牌参数。

3. 设置变频器内部控制参数

通过 ⏎ 与 ▲、▼ 按键，将面板显示调到"Program"—"Interface/Comm"—"Digital Config"中，设置控制模式选择参数"116-L Option Mode"为 2。

注意：当参数"116-L Option Mode"的值写入变频器后，需要对变频器进行断电操作，重新上电并启动变频器后，其内部控制模式才更新为断电前设置的控制模式值。

通过 ⏎ 与 ▲、▼ 按键，将面板显示调到"Program"—"Interface/Comm"—"Analog Inputs"中，设置参数"97-An In 1 Offset"为-5，参数"98-An In 1 Scale"为 4，参数"182-An In 1 Filler"为 0。

通过 ⏎ 与 ▲、▼ 按键，将面板显示调到"Program"—"Control"—"Drive Logic Select"中，设置参数"17-Logic Options"逻辑选项的第 11 位为 1。

通过 ⏎ 与 ▲、▼ 按键，将面板显示调到"Program"—"Control"—"Feedback Device"中，设置参数"64-Fdbk Device Type"反馈方式为编码器反馈"Encoder"。

通过 ← 与 ↑、↓ 按键,将面板显示调到"Program"—"Control"—"Speed Reference"中,设置参数"32 - Speed Ref 3"为 800 rpm,设置参数"33 - Speed Ref 4"为-600 rpm,设置参数"34 - Speed Ref 5"为 500 rpm,设置参数"35 - Speed Ref 6"为 1 200 rpm,设置参数"36 - Speed Ref 7"为 400 rpm。

4. 变频器运行操作

按上述设置变频器内部控制参数的方法,设置相关参数见表 4 - 13。

表 4 - 13 多段速运行操作参数设置

参数号-参数名	功能	出厂值	设定值
116 - L Option Mode	控制模式	1	2
97 - An In 1 Offset	模拟输入 1 偏移量	0	-5
98 - An In 1 Scale	模拟输入 1 比例	2	4
182 - An In 1 Filler	模拟输入 1 滤波器	0	0
17 - Logic Options	逻辑选项	0001 0000 0000 1000	0001 1000 0000 1000
64 - Fdbk Device Type	反馈方式	Encoderless	Encoder
32 - Speed Ref 3	速度参考 3	0 rpm	800 rpm
33 - Speed Ref 4	速度参考 4	0 rpm	-600 rpm
34 - Speed Ref 5	速度参考 5	0 rpm	500 rpm
35 - Speed Ref 6	速度参考 6	0 rpm	1 200 rpm
36 - Speed Ref 7	速度参考 7	0 rpm	400 rpm

控制模式 2(参数 116=2),TB3 控制接口板各端子功能如图 4.15 所示。

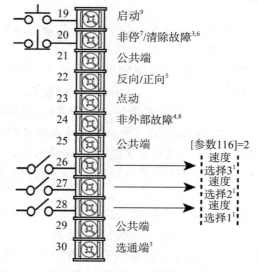

图 4.15 输入模式 2 下 TB3 控制接口板各端子功能

涉及速度选择调节时,可参考变频器内部速度选择表,变频器启动前控制端 20～21 必须提前短接,实验过程中为非停状态。若实验过程中断开 20～21 短接线,则电动机停车。

端子 19 控制电动机启动后,端子 20 为非停状态,端子 26～28 分别为速度选择 3、速度选择 2、速度选择 1 的使能端,根据速度选择表的参考速度定义,实现多段速运行。

第 5 章
PowerFlex 系列变频器实验

§5.1 基于 DeviceNet 的变频器实验

5.1.1 DeviceNet 网络配置及通信

实验目的

(1) 熟悉 DeviceNet 网络架构及总线协议。
(2) 掌握 DeviceNet 网络配置及通信方法。
(3) 掌握 RSLinx、RSLogix 5000 及 RSNetWorx for DeviceNet 软件的使用方法。

实验内容

(1) 完成网络系统模块架构,实现一般通信功能。
(2) 完成 RSNetWorx for DeviceNet 软件的网络配置,获取 dnt 文件。
(3) 将变频器添加到网络中,完成添加设备的通信配置。

实验原理

1. 软件功能描述

(1) RSLinx。

通过 RSLinx 远程配置和浏览网络上的各种硬件,无需预存的配置文件便可浏览整个网络硬件。

(2) RSLogix 5000。

Logix 控制器统一的编程软件。RSLogix 5000 企业版支持 4 种编程语言,包括梯形图、功能块、顺序流程图和结构化文本。

RSLogix 5000 具有如下特点:

① 单一编程软件包支持多种应用项目,通过 RSLogix 5000 可以编写出顺序控制、过程控制、传动控制和运动控制程序。

② Logix 控制器统一的编程环境,一个编程环境可以适合大中小控制系统,用户无须为

不同系统掌握不同的编程软件,节省工程、培训和维护费用。

③ 从网上免费下载最新的固件,在现场就可自己动手为设备进行升级,使原有系统具有新增的功能,保护用户已有投资。

④ 程序编写简单而灵活,指令丰富。

⑤ 基于标记的寻址方式,采用别名。对于一个工程,可以将电气设计和软件编程同时进行,节省开发时间和费用。在编程软件中便可显示趋势图,无需专门软件。

(3) RSNetWorx for DeviceNet。

RSNetWorx 提供对开放设备网供货商协会的 DeviceNet 网络的设计、配置及管理。设置可以在"离线"方式下通过"拖/放"设备图标的操作方式进行,也可以在 RSLinx"在线"扫描 DeviceNet 网络的方式下进行。

2. DeviceNet 电缆结构

DeviceNet 电缆共有 5 根线:电源线(V+ 和 V-)、信号线(CANH 和 CANL)及屏蔽电缆,如图 5.1 所示。

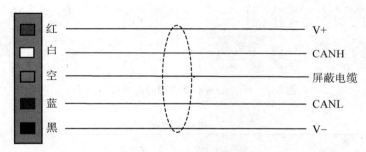

图 5.1　DeviceNet 电缆结构

3. 变频器 PowerFlex 40

PowerFlex 40 的设计结合了应用灵活和控制功能强的优点,以无速度传感器矢量控制和外置 I/O 能力为特征。它具有以下方面的高级特性。

(1) 性能方面。

无速度传感器矢量控制在很宽的速度范围内扩展了高转矩输出,并适应于不同的电机特性。

① 可变的 PWM 允许变频器在低频下输出更大的电流。

② 数字 PID 功能提高了应用的灵活性。

③ 计时器、计数器、基本逻辑和步序逻辑功能可以减少硬件设计成本并简化控制方案。

计时器:通过激活一个被编辑为"计时器启动"的数字输入来启动。

计数器:由变频器控制的继电器或光电耦合输出执行计时功能。

基本逻辑:作为"逻辑输入"编程的数字输入,它的状态控制继电器或光电耦合输出。执行基本的布尔逻辑。

步序逻辑:基于逻辑的步序使用预置的速度设定。每个步序可以按照一个指定的速度、方向和加速/减速曲线进行编程。变频器输出可用于指明正在执行哪个步序。

(2) I/O 方面。

① 两个模拟量输入(一个单极性和一个双极性)与其他变频器 I/O 端子分别隔离。这

些输入可以通过一个数字输入触发。

② 3个固定的和4个完全可编程的数字输入提供了应用的功能性。

③ 一个模拟量输出是通过DIP开关来选择0～10 V或者0～20 mA的。

④ 两个光电耦合输出和一个C型继电器输出通常用于表示变频器、电动机或逻辑量的状态。

（3）通信方面。

① 内置式的DeviceNet集成通信卡可以改善机器的性能。

② 现场安装选件允许以后将独立的变频器扩展到网络上。

③ RSNetWorx在线EDS文件创建功能提供了简便的网络设定。

（4）电源接线端子。

PowerFlex 40变频器的电源接线端子如图5.2所示，接线端子说明如表5-1所示。

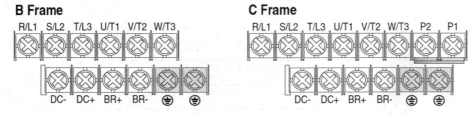

图5.2　PowerFlex 40变频器电源接线端子

表5-1　电源接线端子说明

端子	说　明	
R/L1,S/L2	单相输入	
R/L1,S/L2,T/L3	三相输入	
U/T1	到电动机 U/T1	
V/T2	到电动机 V/T2	交换任何两根电动机导线都可改变方向
W/T3	到电动机 W/T3	
P2,P1	直流母线电抗器连接（仅C Frame变频器）。C Frame变频器在端子P2和P1之间附带跳线。仅在连接直流母线电抗器时才拆除此跳线。如果未连接跳线或电抗器，则变频器不会通电	
DC+,DC-	直流母线连接	
BR+,BR-	动态制动电阻器连接	
	安全接地(PE)	

注意：端子螺丝在运输过程中可能变得松动。向变频器上电之前，请确保所有端子螺丝紧固至推荐的扭矩值。

（5）控制接线端子。

PowerFlex 40变频器的控制接线端子如图5.3所示，启动命令源的说明见表5-2，控制器I/O端子说明见表5-3。

第 5 章 PowerFlex 系列变频器实验

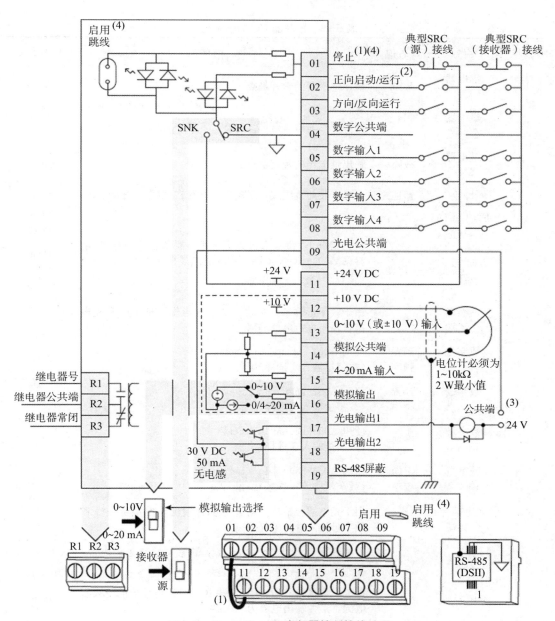

图 5.3　PowerFlex 40 变频器控制接线端子

表 5-2　启动命令源说明

P036 [Start Source]（启动命令源）	停止	I/O 端子 01 停止
键盘	根据 P037	惯性
三线	根据 P037	P037
两线	根据 P037	惯性
RS-485 端口	根据 P037	惯性

表 5-3 控制器 I/O 端子说明

编号	信号	缺省值	说明	参数
R1	继电器常开	故障	输出继电器的常开触点	A055
R2	继电器公共端	—	输出继电器的公共端	
R3	继电器常闭	故障	输出继电器的常闭触点	A055
01	停止	惯性	出厂安装的跳线或常闭输入必须存在,变频器才能启动	P036
02	启动/正向运行	未启动		P036 P037
03	方向/反向运行	未启动	缺省情况下,命令来自数字键盘。要禁止反向操作,请参见 A095[Reverse Disable](反向禁止)	P036 P037 A095
04	数字量公共端	—	用于数字量输入。对数字量输入与模拟量 I/O 和光电耦合输出进行电气隔离	
05	数字量输入 1	预置频率	使用 A051[Digital In 1 Sel] (数字量输入 1 选择)编程	A051
06	数字量输入 2	预置频率	使用 A052[Digital In 2 Sel] (数字量输入 2 选择)编程	A052
07	数字量输入 3	本地	使用 A053[Digital In 3 Sel] (数字量输入 3 选择)编程	A053
08	数字量输入 4	正向点动	使用 A054[Digital In 4 Sel] (数字量输入 4 选择)编程	A054
09	光电耦合公共端	—	用于光电耦合输出。对光电耦合输出与模拟量 I/O 和数字量输入进行电气隔离	
11	+24 V DC	—	参考数字量公共端。用于数字量输入的变频器电源。最大输出电流为 100 mA	
12	+10 V DC	—	参考模拟量公共端。用于 0~10 V 外部电位器的变频器电源。最大输出电流为 15 mA	P038
13	±10 V 输入	未启动	用于外部 0~10 V(单极性)或±10 V(双极性)输入电源(输入阻抗=100 K)或电位器电刷	P038 A051—A054 A123 A132
14	模拟量公共端	—	用于 0~10 V 输入或 4~20 mA 输入。对模拟量输入和输出与数字量 I/O 和光电耦合输出进行电气隔离	
15	4~20 mA 输入	未启动	用于外部 4~20 mA 输入电源(输入阻抗=250)	P038 A051—A054 A132

续表

编号	信号	缺省值	说　　明	参数
16	模拟量输出	输出频率 0～10	缺省模拟量输出为0～10 V。要转换为电流值,可将模拟量输出选择DIP开关更改为0～20 mA。使用A065[Analog Out Sel](模拟量输出选择)编程。可使用A066[Analog Out High](模拟量输出上限)标定最大模拟量值。 最大负载:4～20 mA＝525(10.5 V) 0～10 V＝1 K(10 mA)	A065 A066
17	光电耦合输出1	电动机运行	使用A058[Opto Out 1 Sel] (光电耦合输出1选择)编程	A058 A059 A064
18	光电耦合输出2	达到频率	使用A061[Opto Out 2 Sel] (光电耦合输出2选择)编程	A061 A062 A064
19	RS-485屏蔽	—	使用RS-485(DSI)通信端口时,端子应连接到安全接地(PE)	

重要事项:除非P036[Start Source](启动 命令源)设置为"3-Wire"(三线)或"Momt FWD/REV"(瞬时正向/反向)控制,否则I/O端子01总是设为惯性停止输入方式。在三线控制器中,I/O端子01由P037[Stop Mode](停车模式)控制。所有其他停止源都由P037[Stop Mode](停车模式)控制。

注意:变频器附带有跳线,安装在I/O端子01和11之间。将I/O端子01用作停止输入和使能输入时,拆除此跳线。

所示为两线控制器。对于三线控制器,请在I/O端子02上使用瞬时输入进行启动。对I/O端子03使用保持型输入改变方向。

使用电感式负载(如继电器)的光电耦合输出时,请将继电器并联安装一个续流二极管,以防止损坏输出。

如果拆除ENBL跳线,I/O端子01将始终作为硬件使能,无须软件中断,即可执行惯性停车。

PowerFlex 40内置键盘的外观如图5.4所示,各LED指示灯状态说明见表5-4,按键说明见表5-5。

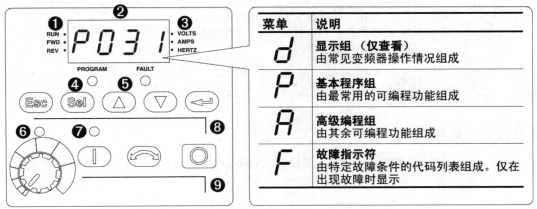

图5.4　PowerFlex 40内置键盘的外观

表 5-4 LED 指示灯状态

编号	LED	LED 状态	说明
1	运行/方向状态	固体红	标识变频器正在运行和命令的电机方向
		闪烁红	变频器接受命令正在改变方向,减速为零时表示实际电动机方向
2	符号显示	固体红	标识参数、参数值或故障代码
		闪烁红	单个数字闪烁标识该参数可被编辑,所有数字闪烁表示出现故障
3	显示单位	固体红	标识当前显示参数单位
4	编程状态	固体红	标识参数值可以被修改
5	故障状态	闪烁红	标识变频器故障
6	电位器状态	固体绿	标识内置键盘上的电位器处于激活状态
7	启动键状态	固体绿	标识内置键盘上的启动键处于激活状态,反向键也已启用,除非由 A095[Reverse Disable](反向禁止)禁止

表 5-5 按键说明表

编号	键	名称	说明
8	Esc	退出	在编程菜单中后退一步。取消参数值的更改并退出程序模式
	Sel	选择	在编程菜单中前进一步。查看参数值时选择一数字
	△▽	向上箭头 向下箭头	在组和参数间滚动。增加/减少闪烁数字的值。在选择 P038[Speed Reference](速度基准值)时用于控制 IP66,NEMA/ULType 4X 等级变频器的速度
	↵	输入	在编程菜单中前进一步。保存对参数值的更改
9	(电位器)	电位器	用于控制变频器的速度。缺省为启用状态。由参数 P038[Speed Reference](速度基准值)控制
	(启动)	启动	用于启动变频器。缺省为启用状态。由参数 P036[Start Source](启动命令源)控制
	(反向)	反向	用于使变频器反向。缺省为启用状态。由参数 P036[Start Source](启动命令源)和 A095[Reverse Disable](反向禁止)控制
	(停止)	停止	用于停止变频器或清除故障。该键始终为启用状态。由参数 P037[Stop Mode](停车模式)控制

实验设备与材料

电源;Logix 5561 CPU;DviceNet 模块;设备网电缆;10 槽机架;PowerFlex 40 变频器。

实验步骤

1. 扫描检测变频器

（1）配置 RSLinx 驱动程序。

（2）配置 DeviceNet 网络。

在开始菜单中查找并打开 RSNetWorx for DeviceNet 软件，如图 5.5 所示。

图 5.5　查找并打开 RSNetWorx for DeviceNet 软件

新建工程文件，并点击菜单栏"NetWork"—"Online"，如图 5.6 所示。

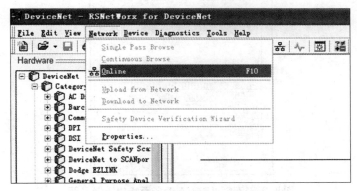

图 5.6　RSNetWorx for DeviceNet 网络上线

启动 RSNetWorx for DeviceNet 后，浏览 DeviceNet 网络设备，如图 5.7 所示。

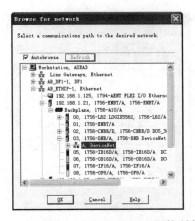

图 5.7　RSNetWorx for DeviceNet 网络浏览界面

点击"OK"按钮,可看到如图 5.8 所示的 DeviceNet 网络上的设备信息。

图 5.8　DeviceNet 网络上的设备

由图 5.8 可见,DeviceNet 网中 02 号节点是变频器 PowerFlex 40。

2. 配置变频器 I/O 参数

(1) 配置变频器的 I/O 参数。

双击图 5.8 中的"1756 - DNB/A"设备网络扫描器图标,点击"Input"选项,弹出如图 5.9 所示的界面。

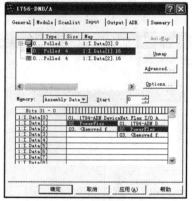

图 5.9　1756 - DNB/A 设备网络扫描器界面

由图 5.9 可见,02 号节点变频器的输入数据在扫描器的 1 和 2 号字节输入数据表中。将其输入数据配置到扫描器的 2 号字节输入数据表中,最终结果如图 5.10 所示。

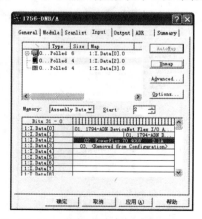

图 5.10　变频器的输入数据界面

依据类似的步骤将变频器的输出数据配置到扫描器的 2 号字节输出数据表中,最终结果如图 5.11 所示。

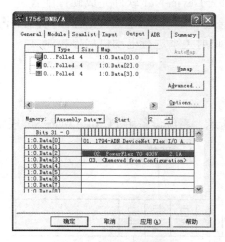

图 5.11 输出数据界面

(2) 保存 dnt 文件。

配置完 I/O 数据后,点击工具栏上的"Save"按钮,保存 dnt 文件,如图 5.12 所示。

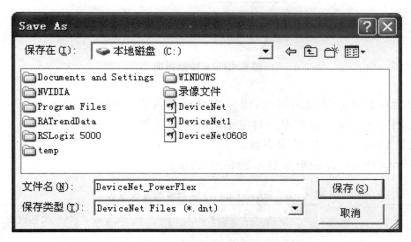

图 5.12 保存 dnt 文件对话框

5.1.2 基于 DeviceNet 控制电动机启停实验

实验目的

(1) 熟悉 DeviceNet 总线协议及网络配置方法。
(2) 掌握 DeviceNet 总线硬件组态及软件组态方法。
(3) 熟悉变频器的使用方法及了解参数的意义。

实验内容

(1) 利用 RSNetWorx for DeviceNet 软件完成硬件配置。
(2) 编写梯形图,实现通过 DeviceNet 控制电动机的启动、停止。
(3) 利用变频器参数的映射关系进行电动机启停控制。

实验原理

1. 控制系统结构

系统结构图如图 5.13 所示。系统设备由 PC 机、DeviceNet 模块、变频器和电动机组成。

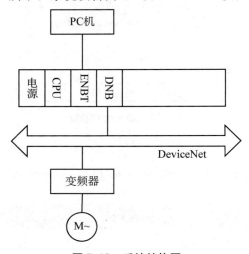

图 5.13 系统结构图

PC 机负责状态监测及发出指令,由以太网模块及设备网模块接入 DeviceNet 总线网络,通过设备网完成变频器的配置及对电动机的控制操作。

2. 变频器 PowerFlex 40 映射参数

变频器 PowerFlex 40 映射参数如表 5-6 所示。

表 5-6 PowerFlex 40 映射参数及含义说明

序号	映射 I/O	参数含义
1	Local:3:O CommandRegister	变频器激活
2	Local:3:O Data[0].1	启动
3	Local:3:O Data[0].0	停止
4	Local:3:O Data[0].3	清除错误
5	Local:3:O Data[0].16-31	设置频率
6	Local:3:I Data[0].16-31	读取频率

注:Local:3:O 中 Local 表示与这些标签相关的模块与控制器位于同一机架中,两个冒号之间的数字代表模块的槽号;例如模块插在 3 槽。紧跟后面冒号显示的字符(如 C,I,O)代表数据是 Configuration,Input 还是 Output 数据。所有模块都有这 3 种数据类型。

3. DeviceNet 通信适配器 22-COMM-D 设置规则

通过 22-COMM-D 通信卡上 DIP 拨码开关设置节点地址和数据传送速率，设置规则如表 5-7 所示，并选择单变频器模式或多变频器模式。

表 5-7 拨码开关设置规则表

拨码开关	说明	缺省值	
SW1	节点地址最低位	1	节点地址 63
SW2	节点地址位 1	1	
SW3	节点地址位 2	1	
SW4	节点地址位 3	1	
SW5	节点地址位 4	1	
SW6	节点地址最高位	1	
SW7	数据速率最低位	1	自动检测速率
SW8	数据速率最高位	1	

4. 电动机启动、停止曲线

被控电动机的转速与时间关系（即输出特性曲线）如图 5.14 所示。

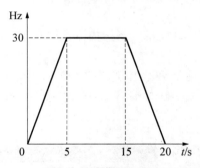

图 5.14 电动机输出曲线

实验设备与材料

电源；Logix 5561 CPU；DviceNet 模块 1756DNB；变频器 PowerFlex 40；设备网所需细缆；10 槽机架；电机。

实验步骤

按照"DeviceNet 网络配置及通信"实验创建配置文件。
按如下步骤编辑控制程序。

1. 创建 RSLogix 5000 工程项目

在 Windows 开始菜单中选择"程序"，选择"Rockwell Software"，再点击 RSLogix 5000 Enterprise Series，选择 RSLogix 5000 即启动了该编程软件。RSLogix 5000 软件界面如图

5.15 所示。打开该界面的文件(file)菜单,选择"New"或点击快捷图标 ,在 RSLogix 5000 软件内创建一个新的工程项目。在随即弹出的如图 5.16 所示的控制器对话框中选择处理器型号(Type——选择 1756 - L61)、版本号(选择 Revision)、名称(Name——自定义)、机架(Chassis——1756 - A10)、槽号(Slot——选择"0"处理器在框架中的实际位置)和所创建的工程项目的位置(Create——选择存盘位置),配置完成后点击"OK"按钮。

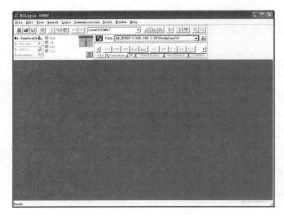

图 5.15　RSLogix 5000 软件界面

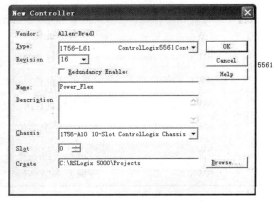

图 5.16　控制器 1756 对话框

2. 配置 I/O 模块

本实验中只用到 1756 - DNB 设备网络扫描器模块和 1756 - IB16D 数字量输入模块,因此只对这两个模块进行配置。

在图 5.17 所示的工程目录列表中选择"I/O Configuration",点击鼠标右键,选择"New Module",弹出 I/O 配置对话框,只勾选"Communication"选项,对话框如图 5.18 所示。

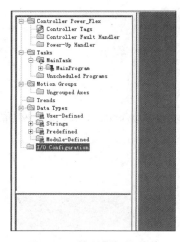

图 5.17　控制器标签界面

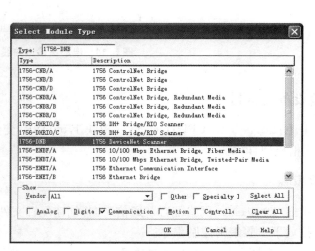

图 5.18　1756 - DNB 对话框

在图 5.18 所示对话框的设备列表中选择 1756 - DNB 设备网络扫描器模块后,点击 "OK" 按钮,弹出如图 5.19 所示的模块配置对话框。

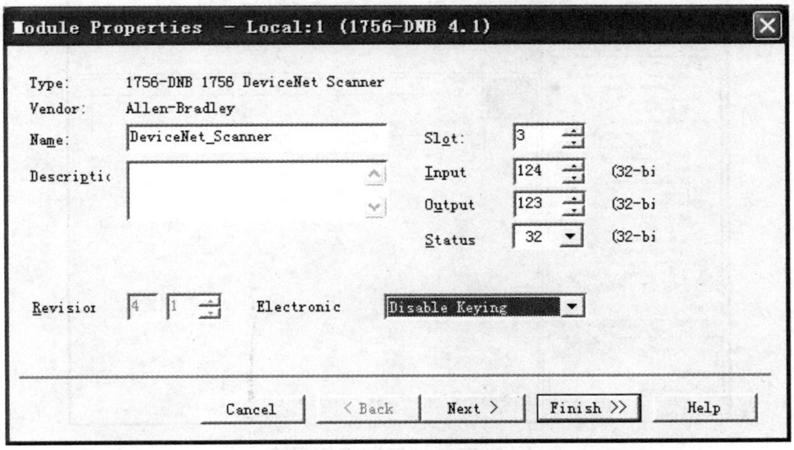

图 5.19 模块配置对话框

在图 5.19 所示对话框中定义模块名称 "Name—可以自定义",按照 1756 - DNB 模块在框架中的实际位置选择槽号(Slot)(注意起始槽号是 0,本实验装置中 1756 - DNB 模块都插在 3 号槽),配置完毕后,点击 "Next" 按钮,直至出现如图 5.20 所示对话框。

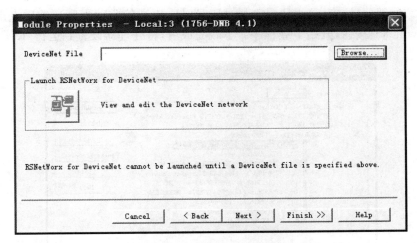

图 5.20 读取 dnt 文件界面

在图 5.20 中点击 "Browse" 按钮,载入第二步中保存的 dnt 文件,完成 1756 - DNB 模块的配置。

3. 编制 Controllogix 控制程序

打开 "Tasks—MainTask—MainProgram" 前面的 "+" 号,在下拉列表中选择 "MainRoutine" 双击,在右侧显示如图 5.21 所示的梯形图编辑界面。

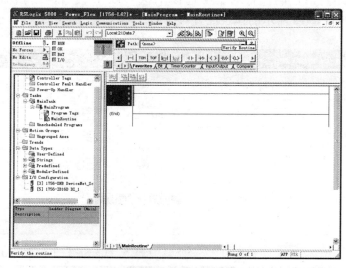

图 5.21 梯形图编辑界面

4. 下载并监视程序运行

程序通过校验后(即梯级左侧没有 e),即可下载到处理器中运行程序了。

(1) 下载程序。

点击图 5.22 中的"AB_ETH-1 Ethernet"前的"+"号,在下拉列表中选择"1756-L62 LOGIX5562"处理器(注意:这里还是选择 AB_ETH-1 通信驱动程序),然后点击右侧的下载程序"Download"按钮,开始下载程序。下载完成后,将处理器设置到运行"RUN"状态。

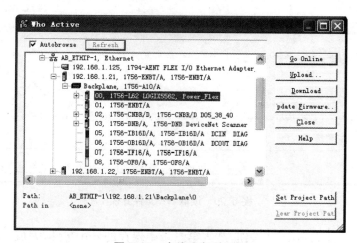

图 5.22 在线设备对话框

(2) 监控程序运行结果。

① 按下启动按钮,观察变频器是否启动。

② 按下停止按钮,观察变频器是否停止运行。

5.1.3 基于 DeviceNet 的变频调速实验

实验目的

(1) 熟悉 DeviceNet 总线的硬件组态及软件组态方法。
(2) 掌握变频器控制电动机的调速方法,并实现 DeviceNet 的网络控制。

实验内容

(1) 利用 RSNetWorx for DeviceNet 软件完成控制系统的优化配置。
(2) 通过对设备网的组态,了解变频器内部的 I/O 点标签的作用。
(3) 编写梯形图,实现 DeviceNet 控制电动机的速度调节。

实验原理

1. 控制系统构成

控制系统由 PC 编程客户端、变频器和电动机组成,如图 5.23 所示。设备连接及实物图如图 5.24 所示。

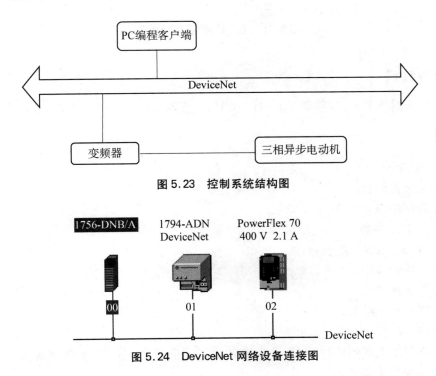

图 5.23 控制系统结构图

图 5.24 DeviceNet 网络设备连接图

2. 运行流程

该实验完成电动机的速度调节,典型的转速变化曲线如图 5.25 所示。
(1) 按下启动按钮,清除变频器错误并启动设备。
(2) 电动机进入高速运转状态,速度值为 A rpm,对应变频器输出频率为 50.0 Hz,同时启动定时器 T1。

(3) 定时器 T1 定时时间到,电动机进入低速运转状态,速度值为 B rpm,对应变频器输出频率为 25.0 Hz,同时启动定时器 T2。

(4) 定时器 T2 定时时间到,电动机进入中速运转状态,速度值为 C rpm,对应变频器输出频率为 40.0 Hz,同时启动定时器 T3。

(5) T3 定时时间到,变频器停止运行,电动机自由停车,变频器输出频率为 0 Hz,自动调速过程结束。

(6) 按下手动停止按钮,自动调速过程强制结束。

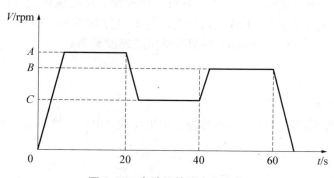

图 5.25　电动机转速变化曲线

实验设备与材料

DeviceNet 网络平台;变频器 PowerFlex 40;编程器;DeviceNet 扫描器;软件环境等。

实验步骤

实验步骤如下:
(1) 网络硬件配置。
(2) 配置网络驱动程序。
(3) 配置 I/O 参数。
(4) 运行 DeviceNet 组态软件。
(5) 编写控制程序。
(6) 程序下载。
(7) 调试过程及结果。

1. 调试过程

(1) 先将程序下载到控制器中,只连接启动与停止开关,先不与变频器相连接,以免输出电压不正确导致变频器出错。

(2) 按下启动按钮,看是否按照规定曲线运行,如果运行正确则证明控制部分调试成功。

(3) 连接输出点与变频器的输入点,并且调试好变频器的参数设置,最后把变频器的输出与电机接好。

(4) 最后按下启动按钮,电机正常运行,并且按照给定的时间函数运行。显示的最大频率是 50 Hz。

2. 调试结果

系统按照给定的时间函数连续运行,软件监测变频器输出频率变化曲线,并验证系统设计是否符合设计要求。

3. 调试方法举例

通过趋势图观测电机的转速变化,如图 5.26 所示,在"Trends"标签右键选择新建"New Trend"。

进入一般设置界面,如图 5.27 所示,设置标题名称"Name"与采样周期"Sample Period",然后点击"下一步"。

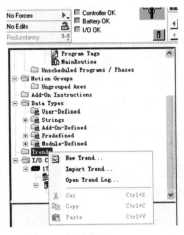

图 5.26　新建趋势图

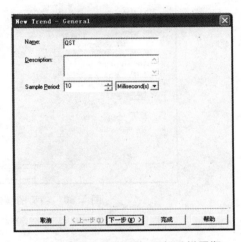

图 5.27　设置趋势图标题与采样周期

进入趋势图的添加/配置标签页面,如图 5.28 所示。添加可用标签"Available Tags",选中需要添加的标签,点击"Add",如图 5.29 所示。在"Tags To Trend"中显示已经添加的标签"MainProgram\EE",点击"完成"。

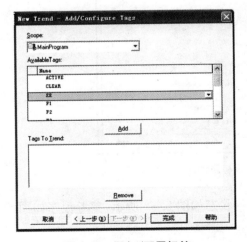

图 5.28　添加/配置标签

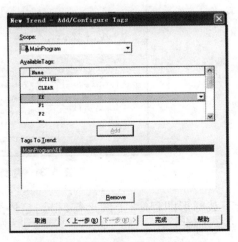

图 5.29　显示已添加/配置的标签

最后需要对趋势图的横纵坐标进行属性设置。如图 5.30 所示，在"RSTrendX 属性"界面中，首先对趋势图的横坐标属性"X-Axis"进行设置，设置横坐标的时间跨度"Time span"，将其改为"1 minute"。然后对趋势图的纵坐标属性"Y-Axis"设置，如图 5.31 所示，在自定义标签栏"Custom"中设置最小值"Minimum value"为"0"与最大值"Maximum value"为"600"，然后点击"确定"。

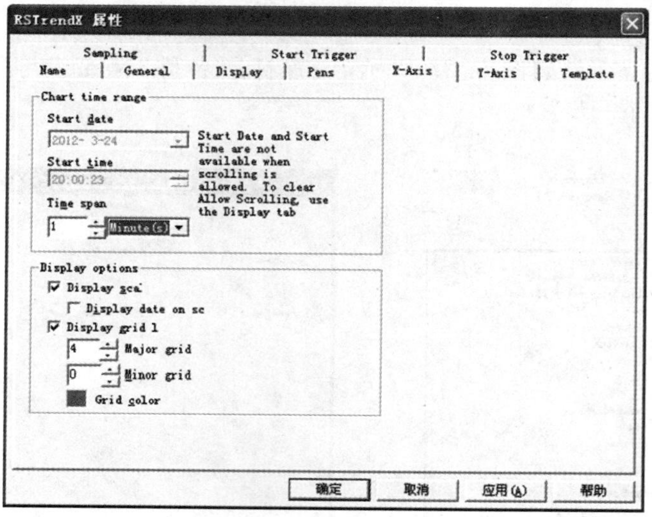

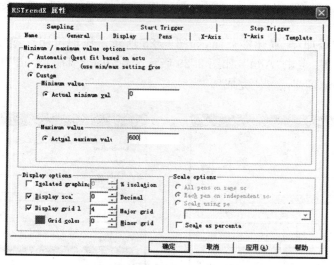

图 5.30　趋势图横坐标属性设置

图 5.31　趋势图纵坐标属性设置

下载梯形图程序，上线运行，得到趋势图运行状态如图 5.32 所示，参考程序如图 5.33 (a)—(c)所示。

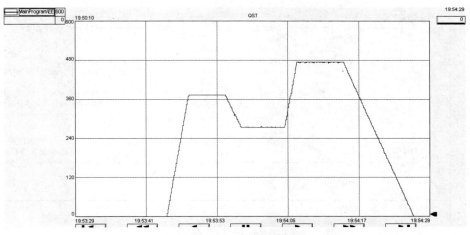

图 5.32 趋势图运行状态

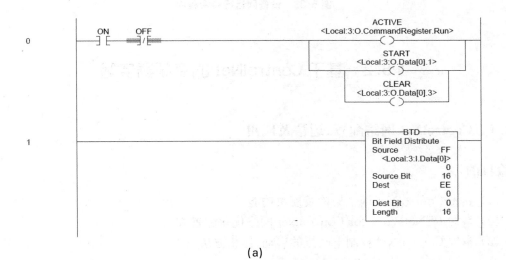

(a)

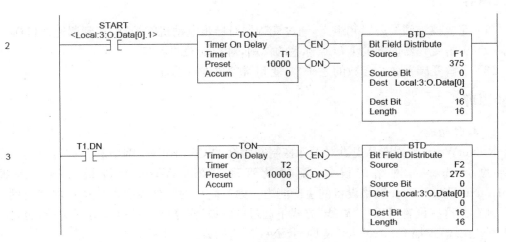

(b)

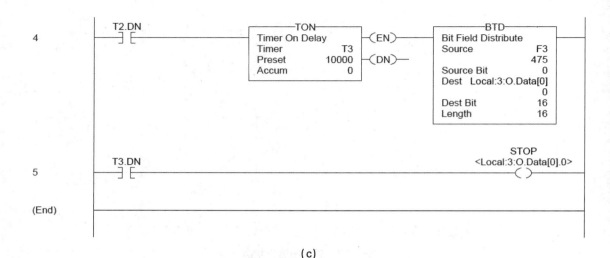

(c)

图 5.33 设备网运行参考程序

§5.2 基于 ControlNet 的变频器实验

5.2.1 ControlNet 网络配置、通信及应用

实验目的

(1) 熟悉 ControlNet 网络架构及模块功能。
(2) 掌握 RSNetWorx for ControlNet 网络优化配置方法。
(3) 掌握 ControlNet 控制电动机的启动、停止方法。

实验内容

(1) 完成控制网的配置和组态,通过变频器对异步电动机的控制,达到调速的目标。
(2) 了解变频器内部的 I/O 点标签的作用,对控制网的组态。
(3) 了解变频器各个参数的意义,对变频器参数进行设置。

实验原理

1. 软件功能

RSNetWorx 系列产品提供对国际控制网协会的 ControlNet 网络和开放设备网供货商协会的 DeviceNet 网络的设计、配置及管理。RSNetWorx 允许最大限度地提高 ControlNet/DeviceNet 网络设备的生产能力。通过简单的软件界面迅速地对网络上的设备进行设置。这些设置可以在"离线"方式下通过"拖/放"设备图标的操作方式进行,也可以在 RSLinx "在线" 扫描 ControlNet 或 DeviceNet 网络的方式下进行。

RSNetWorx 有如下功能:

(1) 充分利用"生产者/消费者"通信模式信息传递的优越性,定义网络上设备的输入/输出数据,便于设备之间相互通信。

(2) 单键式操作实现整个网络配置的上传/下载。

(3) 网络时序排定和带宽计算。

(4) 深层次浏览。

(5) 鼠标点击式配置。

(6) 丰富的设备资源库。

(7) 配置控制器与 I/O 设备之间的关系。

2. 控制系统结构

实验系统结构如图 5.34 所示,系统由 PC 机、以太网模块、设备网模块、变频器及电动机等组成。

PC 机负责状态监测及发出指令,由以太网模块及设备网模块接入 ControlNet 总线网络,通过控制网完成变频器的配置及对电机的控制操作。

3. 变频器映射参数

PowerFlex 40 映射参数及其含义见表 5-8。

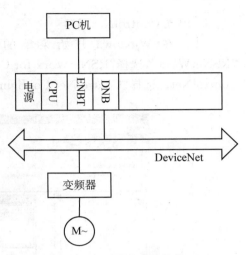

图 5.34　实验系统结构

表 5-8　PowerFlex 40 映射参数及其含义说明

序号	映射 I/O	参数含义
1	PF40:O.Start	变频器启动
2	PF40:O.Stop	停止
3	PF40:O.FreCommond	频率设定
4	PF40:I.OutFreq	读取频率

4. 电动机启动、停止曲线

该实验通过变频器对控制电动机转速进行控制,要求实现转速变化如图 5.35 所示。

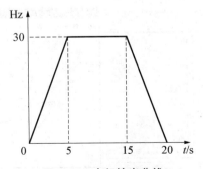

图 5.35　电机输出曲线

实验设备与材料

电源;PLC;ControlNet 模块;变频器 PowerFlex 40;同轴电缆;10 槽机架;以太网模块。

实验步骤

1. 配置 ControlNet 网络

(1) 在 Windows 开始菜单的程序中找到"Rockwell Software",然后选择"RSNetWorx",选择"RSNetWorx for ControlNet",打开网络组态工具软件 RSNetWorx for ControlNet。选择"ControlNet Configuration",配置一个 ControlNet 网络(图 5.36)。

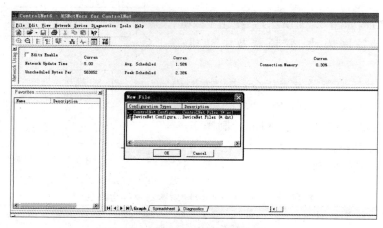

图 5.36　ControlNet 网络配置界面(1)

(2) 点击"Online",通过 AB-ETHIP-1,EtherNet 访问 ContolLogix 背板,选择在网络上的 IP 地址,192.168.1.53 或者 192.168.1.2,然后选择"1756-CNB/E",选择其 A 通道连接的 ControlNet 网络,读取网络上设备信息(图 5.37)。

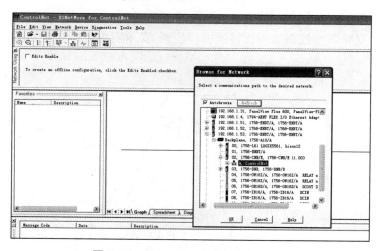

图 5.37　ControlNet 网络配置界面(2)

(3) 点击"OK",如图 5.38 所示。

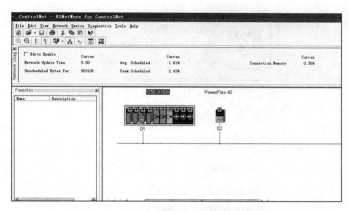

图 5.38　ControlNet 网络配置界面(3)

(4) 网络信息栏中显示了当前的网络状态和参数,在 Edit Enable 前的方框内打勾(出现提示窗口的话按照默认操作),如图 5.39 所示。

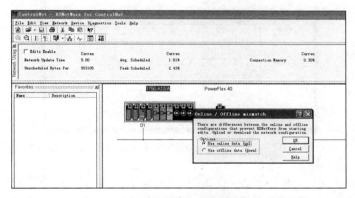

图 5.39　ControlNet 网络配置界面(4)

(5) 点击"OK",如图 5.40 所示。

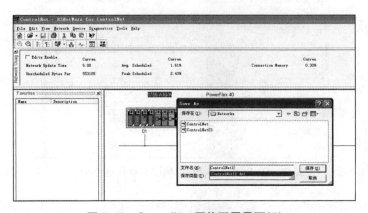

图 5.40　ControlNet 网络配置界面(5)

（6）点击"保存"，即可生成如图 5.41 所示的文件图标。

图 5.41　ControlNet 网络配置文件图标

2. 编程文件的建立

（1）在开始菜单的程序中找到"Rockwell Software"，在"RSLogix 5000 Enterprise series"中打开"RSLogix 5000"，进入 ControlLogix 编程环境。从"File"菜单或常用工具栏中选择"New"新建一个控制项目，输入处理器名称"Name：bianpinqi"，如图 5.42 所示。

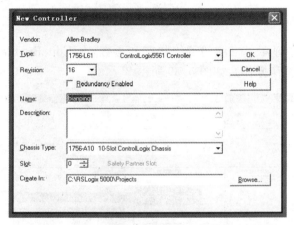

图 5.42　ControlLogix 5561 编程文件的建立对话框

（2）在左侧的项目管理栏下方，右击"I/O Configuration"，选择"New Module"，添加"Communication"下拉菜单下的"1756 - CNB/E ControlNet Brige"通信模块，用于将 ControlLogix 控制器连接到 ControlNet 网络上，如图 5.43 所示。

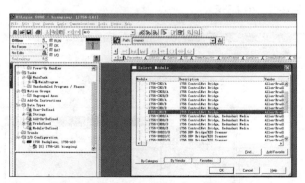

图 5.43　ControlLogix 5561 编程文件的建立界面(1)

（3）在"New Module"窗口中输入模块名称"Name：commu"，选择 CNB 所在槽 2，然后点击"OK"，如图 5.44 所示。

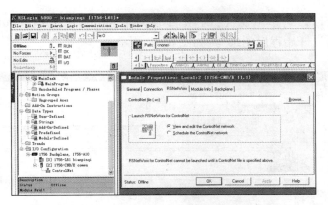

图 5.44　ControlLogix 5561 编程文件的建立界面(2)

（4）点击"RSNetWorx"，然后点击"Browse"，添加完成组态生成的文件，如图 5.45 所示。

图 5.45　ControlLogix 5561 编程文件的建立界面(3)

（5）点击"OK"，完成 ControlNet 的配置。

（6）在 RSLogix 5000 左侧的项目管理栏下方，右击"ControlNet"，选择"New Module"，添加"Drives"下拉菜单下的"PowerFlex 40 - C"，点击"OK"，在"New Module"菜单下的 Name 填写"flex40"，点击"OK"，完成变频器的添加，如图 5.46 所示。

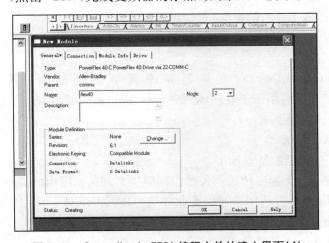

图 5.46　ControlLogix 5561 编程文件的建立界面(4)

（7）点击 RSLogix 5000 左侧的项目管理栏中的 MainProgram 下拉菜单下的 MainRoutine，就可以开始程序的编写，如图 5.47 所示。

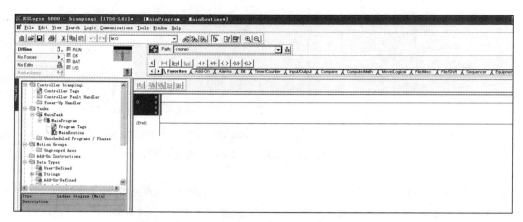

图 5.47　ControlLogix 5561 编程文件的建立界面(5)

完成 ControlLogix 程序的编写，下载程序，运行程序并观察运行结果。

5.2.2　基于 ControlNet 的变频器频率控制

实验目的

（1）熟悉 ControlNet 网络架构及模块功能。
（2）掌握 RSNetWorx for ControlNet 网络优化配置方法。
（3）掌握变频器频率设定及 PLC 控制调节变频器频率的方法。

实验内容

（1）通过对控制网的配置和组态，对变频器的频率进行控制。
（2）通过对控制网的组态，分析和了解各 I/O 点对应的作用，对其进行编程。
（3）熟悉变频器各个参数的意义，对变频器参数进行设置。

实验原理

该实验通过 ControlNet 对变频器频率调节进行控制。其中，变频器采用 PowerFlex 70，它是一款应用于工业现场的变频器，通过高效、灵活的包装及安装方式，可以很好满足客户的各种应用要求。

1. 变频器 PowerFlex 70

PowerFlex 70 变频器的特点如下：

① 电压/频率控制，无速度传感器矢量控制提供宽调速范围内的高转矩控制。

② IP20/NEMA 1 外壳适用于通用工业环境。IP66/NEMA 4X/12 外壳适用于包括高压水喷溅、腐蚀和粉尘等恶劣环境。

③ 法兰安装方式可将变频器的散热片安装在机柜壳后，避免柜内存在发热源。

④ 节省空间的硬件设计包括内置制动单元、制动电阻、EMC 滤波器和通信模块。

⑤ 内置 EMC 滤波器和输出共模磁心提供一体化的集成解决方案,能够满足全球 EMC 电磁兼容性的要求。

⑥ 内置通信选件包括 DeviceNet,ControlNet,EtherNet/IP 和多种其他厂商的开放式通信网络适配卡。

通过 ControlNet 对变频器 PowerFlex 70 频率进行调节控制。同时 PowerFlex 70 也配有独立的操作面板,如图 5.48 所示。

操作面板上的按键定义如表 5-9 所示。

图 5.48　PowerFlex 70 操作面板

表 5-9　操作面板上的按键定义

编程和显示按键		控制按键	
Esc	退出键	↻	正反转
Sel	选择键	∧	加速
▲	向上翻	∨	减速
▼	向下翻	I	启动
↵	回车键	Jog	点动
ALT	ALT 键	O	停止

2. 变频器 PowerFlex 70 映射参数

变频器 PowerFlex 70 映射参数及其含义见表 5-10。

表 5-10　PowerFlex 70 映射参数及含义说明

序号	映射 I/O	参数含义
1	PowerFlex:O. Start	变频器启动
2	PowerFlex:O. Stop	停止
3	PowerFlex:O. ClearFault	清除故障
4	PowerFlex:O. FreCommandedFreq	频率设定
5	PowerFlex:I. OutFreq	读取频率

实验设备与材料

电源;PLC;ControlNet 模块;变频器;控制网所需同轴电缆;10 槽机架。

实验步骤

(1) 双击 RSLogix 5000 图标,依次点击菜单"文件"—"新建"。New Controller(新建控制器)对话框如图 5.49 所示。

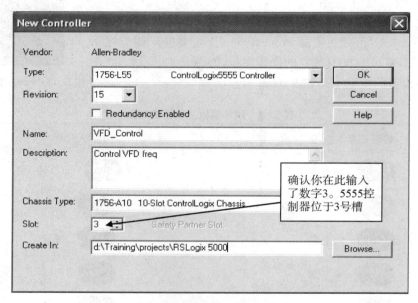

图 5.49　新建控制器对话框

(2) 添加本地机架 CNB 通信模块。鼠标左键点击浏览窗口的"I/O 配置"(位于左边窗口的底部)。然后按鼠标右键,并选择"新建 Module…",在如图 5.50 所示画面中选择"1756 - CNB/D"。

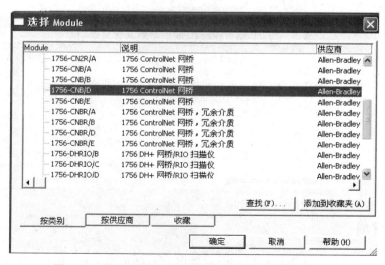

图 5.50　添加 1756 - CNB/D ControlNet 通信模块界面

（3）1756-CNB 通信模块位于第 1 号槽（确认一下），并且在 ControlNet 上的节点地址是 1（观察自己的网络节点地址），因此按图 5.51 所示内容填写。

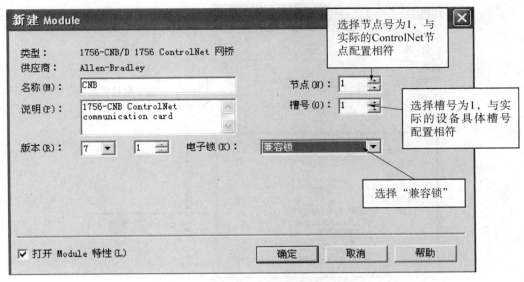

图 5.51　设定 1756-CNB 参数对话框

电子锁允许在 online（联机）之前确定一个物理模块与软件组态之间达到何种匹配程度。这种特性可以避免将错误的模块插入错误的槽中。

（4）接下来添加远程 PowerFlex 70 变频器。在浏览窗口，鼠标右键点击刚添加的 1756-CNB/D ControlNet 通信模块（此模块的名称为 CNB 模块，位于左边浏览窗口的底部），从快捷菜单中选择"新建 Module…"。在"选择 Module"对话框中，单击文件夹"其他"前的"+"，显示此文件下的所有设备类型，从中选择"PowerFlex 70-C"，如图 5.52 所示，选中之后，单击按钮确定。

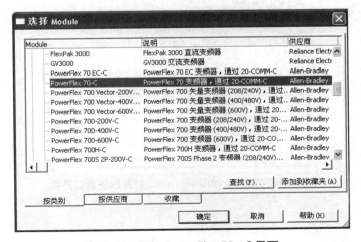

图 5.52　添加 PowerFlex 70-C 界面

(5) 接下来回到主画面，PowerFlex 70 变频器已经添加好，如图 5.53 所示。

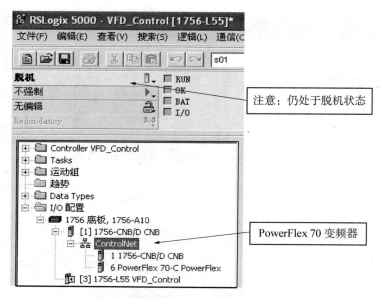

图 5.53　PowerFlex 70-C 的添加后界面

(6) 双击"Controller Tags"，如图 5.54 所示，观察由 RSLogix 5000 编程软件自动生成的 PowerFlex 70 对象数据模型，如图 5.55 所示。

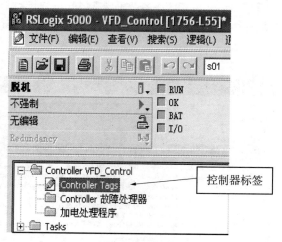

图 5.54　双击"Controller Tags"，打开标签窗口界面

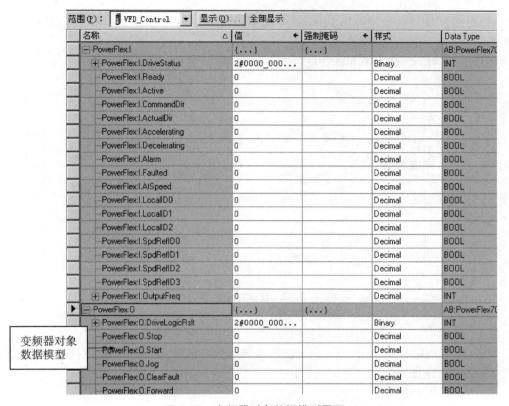

图 5.55　变频器对象数据模型界面

(7) 将鼠标移到"Tasks"文件夹下面的"MainRoutine"。
(8) 按鼠标右键,从快捷菜单上选择打开,出现如图 5.56 所示界面。

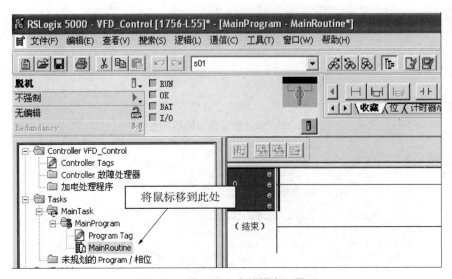

图 5.56　梯形图程序编辑窗口界面

(9) 在指令工具栏上找到相应的指令,点击后,它就出现在阶梯的相应位置。

注意:也可以将其拖到阶梯上,或者左键双击"e"标记,然后在弹出的窗口中输入指令,或者按 Insert 键,输入指令。

如图 5.57 所示,输入梯形图程序。

图 5.57 梯形图程序界面

注意:梯级有错误,因为采用别名编程,没有创建每一个标签。

(10) 为梯形图程序中使用的每一变量名创建相应的标签,输入别名对应的 I/O 地址。添加完毕后,出现如图 5.58 所示的画面。

图 5.58 标签创建好后的程序界面

(11) 确信处于"监控 Tags"状态,在 Freq 的值一栏中输入 5000,如图 5.59 所示。

图 5.59 "监控 Tags"状态界面

（12）双击 MainRoutine，可以看到 Freq 的数值已经显示在 MOV 指令中，如图 5.60 所示。

图 5.60　标签值已更改的程序显示界面

（13）在主菜单上，依次点击"文件"—"保存"，或直接点击工具栏上的图标 ![save] 保存程序，如图 5.61 所示。

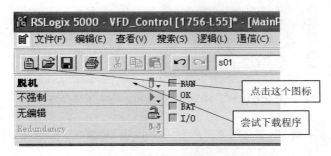

图 5.61　保存并下载程序界面

（14）下载程序到控制器。依次点击菜单上的"通信"—"活动项"，出现如图 5.62 所示的界面：直到选择了 03 号槽的 L55 处理器。选中"设置项目路径"，然后选择"下载"。

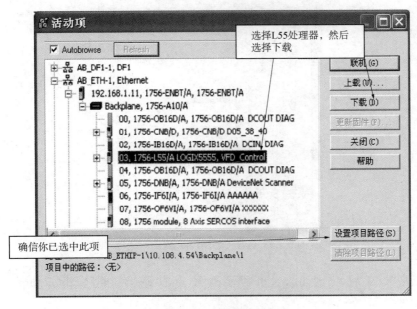

图 5.62　程序下载对话框

(15) 双击启动"RSNetWorx for CntrolNet"。

(16) 进入如图 5.63 所示的界面。

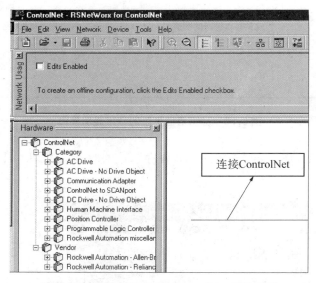

图 5.63 RSNetWorx for ControlNet 网络界面

(17) 点击"联机"按钮，选择以太网的驱动，上到背板，找到 CNB 网卡，最后进入 ControlNet，如图 5.64 所示。

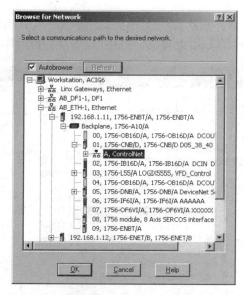

图 5.64 连接 ControlNet 网络对话框

(18) 点击"OK"后，出现如图 5.65 所示画面，点击"编辑使能"(Edit Enabled)。

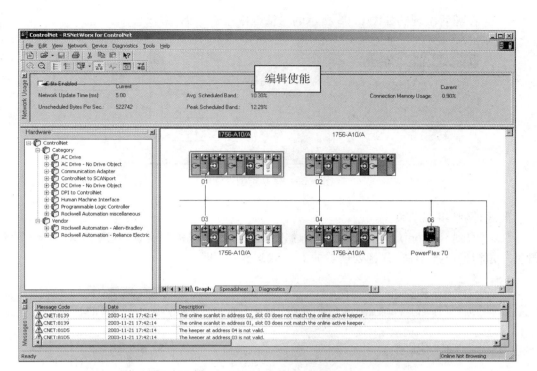

图 5.65　联机 ControlNet 网络界面

（19）单击菜单 Network 的"Properties"项进行网络参数设定，如图 5.66 所示。

（20）点击后，出现了 ControlNet 网络参数设定对话框，并按照如图 5.67 所示改变"Max Scheduled Address"为"6"。改变"Max Unscheduled Address"为"11"。最后单击按钮"OK"，并保存设置信息。

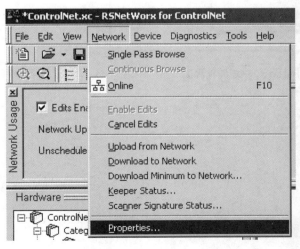

图 5.66　启动 ControlNet 网络参数设定窗口界面

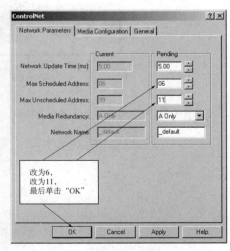

图 5.67　ControlNet 网络参数设定对话框

（21）在随后出现的如图 5.68 所示的对话框中单击"OK"继续。

图 5.68　保存类型对话框

至此，网络组态结束。

（22）回到 RSLogix 5000 编程界面，如图 5.69 所示，联机并观察，发现 PowerFlex 70 的黄色三角标记已经消失，并且处理器的 I/O 状态显示 OK。

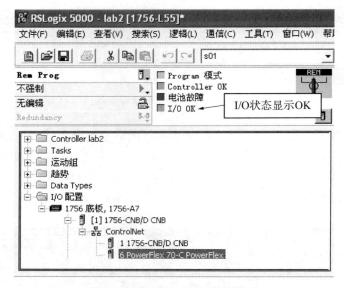

图 5.69　RsLogix 5000 编程界面

现在可以开启变频器和调节频率。

1336 Impact 变频器常用参数

1 Language Select

Use *Language Select* to choose between a primary language and an alternate language. Select:
- 0 to choose the primary language
- 1 to choose the alternate language

Parameter number	1
File:group	none
Parameter type	linkable destination
Display	x
Factory default	0
Minimum value	0
Maximum value	1
Conversion	1 = 1

2 Nameplate HP

Nameplate HP contains the value of the motor horsepower that you entered during the start up routine. This value is typically located on the motor nameplate.

Parameter number	2
File:group	Motor/Inverter:Motor Nameplate
Parameter type	destination
Display	x.x hp
Factory default	30.0 hp
Minimum value	0.2 hp
Maximum value	2000.0 hp
Conversion	10 = 1.0

3 Nameplate RPM

Nameplate RPM contains the value of the motor speed that you entered during the start up routine. This value is typically located on the motor nameplate. This value should not be the synchronous speed of the motor.

Parameter number	3
File:group	Motor/Inverter:Motor Nameplate
Parameter type	destination
Display	x rpm
Factory default	1750 rpm
Minimum value	1 rpm
Maximum value	15000 rpm
Conversion	1 = 1

4 Nameplate Amps

Nameplate Amps contains the value of the current rating of the motor that you entered during the start up routine. This value is typically located on the motor nameplate. The drive uses this information to properly tune to the motor.

Parameter number	4
File:group	Motor/Inverter:Motor Nameplate
Parameter type	destination
Display	x.x amps
Factory default	0.2 amps
Minimum value	0.1 amps
Maximum value	calculated
Conversion	10 = 1.0

5 Nameplate Volts

Nameplate Volts contains the voltage rating of the motor that you entered during the start up routine. This value is typically located on the motor nameplate.

Parameter number	5
File:group	Motor/Inverter:Motor Nameplate
Parameter type	destination
Display	x volt
Factory default	460 volts
Minimum value	75 volts
Maximum value	575 volts
Conversion	1 = 1

6 Nameplate Hz

Nameplate Hz contains the value of the frequency rating of the motor that you entered during the start up routine. This value is typically located on the motor nameplate.

Parameter number	6
File:group	Motor/Inverter:Motor Nameplate
Parameter type	destination
Display	x.x Hz
Factory default	60.0 Hz
Minimum value	1.0 Hz
Maximum value	250.0 Hz
Conversion	10 = 1.0

7 Motor Poles

Motor Poles contains the number of motor poles. The drive calculates this value during the Quick Motor Tune portion of the start up routine.

Note: <u>Encoder PPR</u> Must be greater than 64
\# of Motor Poles

Parameter number	7
File:group	Motor/Inverter:Motor Nameplate
	Motor/Inverter:Motor Constants
Parameter type	destination
Display	x poles
Factory default	4 poles
Minimum value	2 poles
Maximum value	40 poles
Conversion	1 = 1

8	**Encoder PPR**		
	Encoder PPR contains the pulse per revolution rating of the feedback device when you use an encoder to determine motor speed.	Parameter number	8
		File:group	Motor/Inverter:Encoder Data
			Control:Feedback Device
		Parameter type	destination
		Display	x ppr
		Factory default	1024 ppr
	Note: $\text{Encoder PPR} \over \text{\# of Motor Poles}$ Must be greater than 64	Minimum value	calculated
		Maximum value	20000 ppr
		Conversion	1 = 1

9	**Service Factor**		
	Enter the minimum level of current that causes a motor overload (I^2T) trip under continuous operation. Current levels below this value never result in an overload trip. For example, a service factor of 1.15 implies continuous operation up to 115% of nameplate motor current.	Parameter number	9
		File:group	Motor/Inverter:Motor Nameplate
		Parameter type	destination
		Display	x.xx
		Factory default	1.15
		Minimum value	1.00
		Maximum value	2.00
		Conversion	4096 = 1.00

10	**PWM Frequency**		
	Enter the drive carrier frequency in Hz. The drive carrier frequency depends on your application and drive size. The drive carrier frequency affects the audible noise level of your motor.	Parameter number	10
		File:group	Motor/Inverter:Inverter
		Parameter type	destination
		Display	x Hz
		Factory default	4000 Hz
		Minimum value	1000 Hz
		Maximum value	from the drive type
		Conversion	1 = 1

11	**Inverter Amps**		
	Inverter Amps provides the current rating of the inverter. The drive automatically sets *Inverter Amps* at power up.	Parameter number	11
		File:group	Motor/Inverter:Inverter
		Parameter type	source
		Display	x.x amps
		Factory default	not applicable
		Minimum value	0.1 amps
		Maximum value	from drive type
		Conversion	10 = 1.0

12	**Inverter Volts**		
	Inverter Volts is the drive nameplate voltage rating of the inverter. The drive automatically sets *Inverter Volts* at power up.	Parameter number	12
		File:group	Motor/Inverter:Inverter
		Parameter type	source
		Display	x volts
		Factory default	not applicable
		Minimum value	75 volts
		Maximum value	575 volts
		Conversion	1 = 1

14	**Logic Input Sts**		
	Use *Logic Input Sts* to view drive logic operation. If a bit is set (1), that function is enabled. If a bit is clear (0), that function is disabled (not active).	Parameter number	14
		File:group	Monitor:Drive/Inv Status
		Parameter type	source
		Display	bits
		Factory default	not applicable
		Minimum value	00000000.00000000
	The bits are defined as follows:	Maximum value	11111111.11111111
		Conversion	1 = 1

Bit	Description	Bit	Description	Bit	Description
0	**Normal Stop** A ramp stop is selected.	5	**Reverse** A reverse was commanded.	10	**Flux Enable** Flux is enabled.
1	**Start** A start is in progress.	6	**Jog 2** A jog 2 is in progress.	11	**Process Trim** Process trim is enabled.
2	**Jog 1** A jog 1 is in progress.	7	**Cur Lim Stop** A current limit stop is selected.	12	**Speed Ref A**
3	**Clear Fault** A clear fault is in progress.	8	**Coast Stop** A coast stop is selected.	13	**Speed Ref B**
4	**Forward** A forward was commanded.	9	**Spd Ramp Dis** Ramps are disabled.	14	**Speed Ref C**
				15	**Reset Drive** The drive has been commanded to reset.

C	B	A	
0	0	0	No Change
0	0	1	Speed Ref 1
0	1	0	Speed Ref 2
0	1	1	Speed Ref 3
1	0	0	Speed Ref 4
1	0	1	Speed Ref 5
1	1	0	Speed Ref 6
1	1	1	Speed Ref 7

17 Logic Options

Use *Logic Options* to select the options for logic operation of the drive.

If you set bits 1, 2, and 3, the drive performs a coast to stop. For additional information about the stop types and priorities, refer to Appendix B, *Control Block Diagrams*.

The bits are defined as follows:

Parameter number	17
File:group	Control:Drive Logic Select
Parameter type	linkable destination
Display	bits
Factory default	00010000.00001000
Minimum value	00000000.00000000
Maximum value	01111111.11111111
Conversion	1 = 1

Bit	Description	Bit	Description	Bit	Description
0	Reserved Leave 0.	7	Jog Coast 1 selects jog coast. 0 selects regen stop.	12	Coast Stop 2 Set to use a coast to stop. Only used when *L Option Mode* (par. 116) is 3, 13, or 16.
1	Coast Stop 1 Set to use a coast to stop.	8	Start Diag Do diagnostics each time the drive is started.	13	CurLim Stop 2 Set to use a current limit to stop. Only used when *L Option Mode* (par. 116) is 3, 13, or 16.
2	CurLim Stop 1 Set to use a current limit to stop.	9	Pwr Up Start Set to enable the auto start feature on power up if a start is valid.	14	Ramp Stop 2 Set to use a ramp to stop. Only used when *L Option Mode* (par. 116) is 3, 13, or 16.
3	Ramp Stop 1 Set to use a ramp to stop.	10	Reserved Leave 0.	15	Reserved Leave 0.
4–5	Reserved Leave 0.	11	Bipolar Sref 1 selects bipolar reference. 0 selects unipolar reference.		
6	Jog Ramp En Set to enable the jog ramp.				

18 Stop Dwell Time

Use *Stop Dwell Time* to set an adjustable delay time before the drive disables speed and torque regulators when a stop has been initiated.

Parameter number	18
File:group	Control:Drive Logic Select
Parameter type	linkable destination
Display	x.x seconds
Factory default	0.0 seconds
Minimum value	0.0 seconds
Maximum value	10.0 seconds
Conversion	10 = 1.0

19 Zero Speed Tol

Use *Zero Speed Tol* to establish a band around zero speed that is used to determine when the drive considers the motor to be at zero speed. Bit 12 (At Zero Spd) in *Drive/Inv Status* (parameter 15) indicates this.

Parameter number	19
File:group	Control:Drive Logic Select
Parameter type	linkable destination
Display	x.x rpm
Factory default	base motor speed/100 rpm
Minimum value	0.0 rpm
Maximum value	8 x base motor speed rpm
Conversion	4096 = base motor speed

20 Fault Select 1

Use *Fault Select 1* to specify how the drive should handle certain conditions. Each bit within this parameter matches the bit definitions of *Warning Select 1* (parameter 21). If you set bit(s) to 1 in this parameter, the drive reports a fault when that condition occurs. If you clear bit(s) to 0, the drive reports the condition based on *Warning Select 1*.

Parameter number	20
File:group	Fault Setup:Fault Config
Parameter type	linkable destination
Display	bits
Factory default	01111110.00100011
Minimum value	00000000.00000000
Maximum value	01111111.00111111
Conversion	1 = 1

The bits are defined as follows:

Bit	Description	Bit	Description	Bit	Description
0	RidethruTime A bus ridethrough timeout occurred.	6–7	Reserved Leave 0.	12	SP 4 Timeout Loss of communication with SCANport device 4 occurred.
1	Prechrg Time A bus precharge timeout occurred.	8	mA Input A loss of input connection occurred after it was established.	13	SP 5 Timeout Loss of communication with SCANport device 5 occurred.
2	Bus Drop A bus drop of 150 volts occurred.	9	SP 1 Timeout Loss of communication with SCANport device 1 occurred.	14	SP 6 Timeout Loss of communication with SCANport device 6 occurred.
3	Bus Undervlt A bus undervoltage occurred.	10	SP 2 Timeout Loss of communication with SCANport device 2 occurred.	15	SP Error Too many errors on the SCANport communication.
4	Bus Cycles>5 More than 5 ridethroughs occurred in a row.	11	SP 3 Timeout Loss of communication with SCANport device 3 occurred.		
5	Open Circuit Fast flux up current is <50%.				

21	**Warning Select 1**				
	Use *Warning Select 1* to specify how the drive should handle certain conditions. Each bit within this parameter matches the bit definitions of *Fault Select 1* (parameter 20). If you set a bit to 1 and the corresponding bit in *Fault Select 1* is clear (0), the drive reports a warning when that condition occurs. If both corresponding bits in *Fault Select 1* and *Warning Select 1* are 0, the drive ignores the condition when it occurs.			Parameter number File:group Parameter type Display Factory default Minimum value Maximum value Conversion	21 Fault Setup:Fault Config linkable destination bits 00000000.00011100 00000000.00000000 01111111.00111111 1 = 1
	The bits are defined as follows:				

Bit	Description	Bit	Description	Bit	Description
0	**RidethruTime** A bus ridethrough timeout occurred.	6–7	**Reserved** Leave 0.	12	**SP 4 Timeout** Loss of communication with SCANport device 4 occurred.
1	**Prechrg Time** A bus precharge timeout occurred.	8	**mA Input** A loss of input connection occurred after it was established.	13	**SP 5 Timeout** Loss of communication with SCANport device 5 occurred.
2	**Bus Drop** A bus drop of 150 volts occurred.	9	**SP 1 Timeout** Loss of communication with SCANport device 1 occurred.	14	**SP 6 Timeout** Loss of communication with SCANport device 6 occurred.
3	**Bus Undervlt** A bus undervoltage occurred.	10	**SP 2 Timeout** Loss of communication with SCANport device 2 occurred.	15	**SP Error** Too many errors on the SCANport communication.
4	**Bus Cycles>5** More than 5 ridethroughs occurred in a row.	11	**SP 3 Timeout** Loss of communication with SCANport device 3 occurred.		
5	**Open Circuit** Fast flux up current is <50%.				

22	**Fault Select 2**				
	Use *Fault Select 2* to specify how the drive should handle certain conditions. Each bit matches the bit definitions of *Warning Select 2* (parameter 23). If you set a bit to 1, the drive reports a fault when that condition occurs. If you clear a bit to 0, the drive reports the condition based on *Warning Select 2*.			Parameter number File:group Parameter type Display Factory default Minimum value Maximum value Conversion	22 Fault Setup:Fault Config linkable destination bits 10000000.00010001 00000000.00000000 11111111.11111111 1 = 1
	The bits are defined as follows:				

Bit	Description	Bit	Description	Bit	Description
0	**SpdFdbk Loss** A loss of feedback occurred.	5	**Mtr Stall** The motor stalled.	11–12	**Reserved** Leave 0.
1	**InvOvtmp Pnd** An inverter overtemp is pending.	6	**Ext Fault In** The ext input is open.	13	**InvOvld Pend** An inverter overload is pending (IT).
2	**Reserved** Leave 0.	7–8	**Reserved** Leave 0.	14	**Reserved** Leave 0.
3	**MtrOvld Pend** A motor overload is pending (I^2T).	9	**Param Limit** A parameter is out of limits	15	**InvOvld Trip** Inverter overload trip (IT)
4	**MtrOvld Trip** Motor overload trip (I^2T)	10	**Math Limit** A math limit occurred.		

28	**Speed Ref 1 Frac**		
	Use *Speed Ref 1 Frac* to supply the fractional part of the external speed reference 1 when speed reference is selected in *Logic Input Sts* (parameter 14).	Parameter number File:group Parameter type Display Factory default Minimum value Maximum value Conversion	28 none linkable destination x 0 0 65535 1 = 1/2^28 base motor speed

29	**Speed Ref 1**		
	Enter the speed reference that the drive should use when speed reference 1 is selected in *Logic Input Sts* (parameter 14). *Speed Ref 1* supplies the whole number portion of the speed reference. You can use *Speed Ref 1 Frac* (parameter 28) to specify the fractional portion of the speed reference.	Parameter number File:group Parameter type Display Factory default Minimum value Maximum value Conversion	29 Control:Speed Reference linkable destination ±x.x rpm 0.0 rpm -8 x base motor speed rpm +8 x base motor speed rpm 4096 = base motor speed

30	**Speed Scale 1**		
	Enter the gain multiplier used to scale speed reference 1.	Parameter number File:group Parameter type Display Factory default Minimum value Maximum value Conversion	30 Control:Speed Reference linkable destination ±x.xxxx +1.0000 -3.9999 +3.9999 8192 = 1.0000

附录 A　1336 Impact 变频器常用参数

31　Speed Ref 2
Enter the speed reference that the drive should use when speed reference 2 is selected in *Logic Input Sts* (parameter 14).

Parameter number	31
File:group	Control:Speed Reference
Parameter type	linkable destination
Display	±x.x rpm
Factory default	0.0 rpm
Minimum value	-8 x base motor speed rpm
Maximum value	+8 x base motor speed rpm
Conversion	4096 = base motor speed

32　Speed Ref 3
Enter the speed reference that the drive should use when speed reference 3 is selected in *Logic Input Sts* (parameter 14).

Parameter number	32
File:group	Control:Speed Reference
Parameter type	linkable destination
Display	±x.x rpm
Factory default	+0.0 rpm
Minimum value	-8 x base motor speed rpm
Maximum value	+8 x base motor speed rpm
Conversion	4096 = base motor speed

33　Speed Ref 4
Enter the speed reference that the drive should use when speed reference 4 is selected in *Logic Input Sts* (parameter 14).

Parameter number	33
File:group	Control:Speed Reference
Parameter type	linkable destination
Display	±x.x rpm
Factory default	+0.0 rpm
Minimum value	-8 x base motor speed rpm
Maximum value	+8 x base motor speed rpm
Conversion	4096 = base motor speed

34　Speed Ref 5
Enter the speed reference that the drive should use when speed reference 5 is selected in *Logic Input Sts* (parameter 14).

Parameter number	34
File:group	Control:Speed Reference
Parameter type	linkable destination
Display	±x.x rpm
Factory default	+0.0 rpm
Minimum value	-8 x base motor speed rpm
Maximum value	+8 x base motor speed rpm
Conversion	4096 = base motor speed

35　Speed Ref 6
Enter the speed reference that the drive should use when speed reference 6 is selected in *Logic Input Sts* (parameter 14).

Parameter number	35
File:group	Control:Speed Reference
Parameter type	linkable destination
Display	±x.x rpm
Factory default	+0.0 rpm
Minimum value	-8 x base motor speed rpm
Maximum value	+8 x base motor speed rpm
Conversion	4096 = base motor speed

36　Speed Ref 7
Enter the speed reference that the drive should use when speed reference 7 is selected in *Logic Input Sts* (parameter 14).

Parameter number	36
File:group	Control:Speed Reference
Parameter type	linkable destination
Display	±x.x rpm
Factory default	+0.0 rpm
Minimum value	-8 x base motor speed rpm
Maximum value	+8 x base motor speed rpm
Conversion	4096 = base motor speed

37　Speed Scale 7
Enter the gain multiplier used to scale *Speed Ref 7* (parameter 36).

Parameter number	37
File:group	Control:Speed Reference
Parameter type	linkable destination
Display	±x.xxxx
Factory default	+1.0000
Minimum value	-3.9999
Maximum value	+3.9999
Conversion	8192 = 1.0000

38　Jog Speed 1
Enter the speed reference that the drive should use when Jog 1 is selected in *Logic Input Sts* (parameter 14).

Parameter number	38
File:group	Control:Speed Reference
Parameter type	linkable destination
Display	±x.x rpm
Factory default	+100.0 rpm
Minimum value	-8 x base motor speed rpm
Maximum value	+8 x base motor speed rpm
Conversion	4096 = base motor speed

39	**Jog Speed 2**	Parameter number	39
	Enter the speed reference that the drive should use when Jog 2 is selected in *Logic Input Sts* (parameter 14).	File:group	Control:Speed Reference
		Parameter type	linkable destination
		Display	±x.x rpm
		Factory default	+0.0 rpm
		Minimum value	-8 x base motor speed rpm
		Maximum value	+8 x base motor speed rpm
		Conversion	4096 = base motor speed
42	**Accel Time 1**	Parameter number	42
	Enter the length of time for the drive to ramp from 0 rpm to the base speed.	File:group	Control:Accel/Decel
		Parameter type	linkable destination
		Display	x.x seconds
		Factory default	5.0 seconds
		Minimum value	0.0 seconds
		Maximum value	6553.5 seconds
		Conversion	10 = 1.0
43	**Accel Time 2**	Parameter number	43
	Enter the length of time for the drive to ramp from 0 rpm to the base speed. *Accel Time 2* is available only when the value of *L Option Mode* (parameter 116) is 4, 11, or 14.	File:group	Control:Accel/Decel
		Parameter type	linkable destination
		Display	x.x seconds
		Factory default	10.0 seconds
		Minimum value	0.0 seconds
		Maximum value	6553.5 seconds
		Conversion	10 = 1.0
44	**Decel Time 1**	Parameter number	44
	Enter the length of time for the drive to ramp from base speed to 0 rpm. This is used for a ramp stop.	File:group	Control:Accel/Decel
		Parameter type	linkable destination
		Display	x.x seconds
		Factory default	5.0 seconds
		Minimum value	0.0 seconds
		Maximum value	6553.5 seconds
		Conversion	10 = 1.0
45	**Decel Time 2**	Parameter number	45
	Enter the length of time for the drive to ramp from base speed to 0 rpm. This is used for a ramp stop. *Decel Time 2* is available only when the value of *L Option Mode* (parameter 116) is 4, 11, or 14.	File:group	Control:Accel/Decel
		Parameter type	linkable destination
		Display	x.x seconds
		Factory default	10.0 seconds
		Minimum value	0.0 seconds
		Maximum value	6553.5 seconds
		Conversion	10 = 1.0
63	**Scaled Spd Fdbk**	Parameter number	63
	Scaled Spd Fdbk is a scaled version of speed feedback. The inverse of either *Speed Scale 1* (parameter 30) or *Speed Scale 7* (parameter 37) is used.	File:group	Control:Speed Feedback
		Parameter type	source
		Display	±x
		Factory default	not applicable
		Minimum value	-32767
		Maximum value	+32767
		Conversion	1 = 1
64	**Fdbk Device Type**	Parameter number	64

	Use *Fdbk Device Type* to choose the source for motor speed feedback from the following options:
	
	Value Description
	1 **Encoderless**
	  Use this mode if you do not have an encoder.
	2 **Encoder**
	  Use this mode if you do have an encoder.
	3 **Simulator**
	  Use this mode to simulate a motor.
	4 **Encoderless W/Deadband**
	  Use this mode if you do not have an encoder and operation below 1Hz is not required.
	
	Whenever possible, you should use the start up procedure to change the feedback device type because the start up procedure automatically re-adjusts the speed loop gains when you change between encoder and encoderless operation.
		File:group	Control:Feedback Device

		Parameter type	destination
		Display	x
		Factory default	1
		Minimum value	1
		Maximum value	3
		Conversion	1 = 1

附录 A　1336 Impact 变频器常用参数　　

65	**Fdbk Filter Sel**		
	Use *Fdbk Filter Sel* to select the type of feedback filter. You can choose among the following filters:	Parameter number	65
		File:group	Control:Speed Feedback
		Parameter type	linkable destination
		Display	x
		Factory default	0
		Minimum value	0
		Maximum value	4
		Conversion	1 = 1

Value	Description
0	**No Filter** — Use this option if you do not want to filter the feedback.
1	**35/49 rad** — Use a "light" 35/49 radian feedback filter.
2	**20/40 rad** — Use a "heavy" 20/40 radian feedback filter.
3	**Lead/Lag** — Use a single pole lead lag feedback filter. You need to set up *Fdbk Filter Gain* (par. 66) and *Fdbk Filter BW* (par. 67).
4	**Notch** — Use a notch filter. You need to set up *Notch Filtr Freq* (par. 185) and *Notch Filtr Q* (par. 186).

66	**Fdbk Filter Gain**		
	Use *Fdbk Filter Gain* to specify the Kn term of the single pole lead/lag feedback filter.	Parameter number	66
		File:group	Control:Speed Feedback
		Parameter type	linkable destination
		Display	±x.xx
		Factory default	+1.00
		Minimum value	-5.00
		Maximum value	+5.00
		Conversion	256 = 1.00

If KN Is:	Then:
Greater than 1.0	A lead filter is produced.
Less than 1.0	A lag filter is produced.
Equal to 1.0	The feedback filter is disabled.
Equal to 0.0	A simple, low pass filter is produced.

You need to set this parameter if *Fdbk Filter Sel* (parameter 65) is set to 3.

67	**Fdbk Filter BW**		
	Use *Fdbk Filter BW* to establish the breakpoint frequency (in radians) for the speed feedback lead/lag filter. You need to set this parameter if *Fdbk Filter Sel* (parameter 65) is set to 3.	Parameter number	67
		File:group	Control:Speed Feedback
		Parameter type	linkable destination
		Display	x.x radians/second
		Factory default	100.0 radians/second
		Minimum value	0.2 radian/second
		Maximum value	900.0 radians/second
		Conversion	10 = 1.0

79	**DC Brake Current**[1]		
	Enter the percent of motor current to be used for DC braking the motor. To enable DC braking, you need to set bit 9 in *Bus/Brake Opts* (parameter 13).	Parameter number	79
		File:group	Application:DC Braking/Hold
		Parameter type	linkable destination
		Display	x.x%
		Factory default	50.0%
		Minimum value	0.0%
		Maximum value	calculated
		Conversion	4096 = 100.0% current

1　*DC Brake Current* was added in Version 2.xx.

80	**DC Brake Time**[1]		
	Enter the period of time that the DC braking current should be applied after a stop has been commanded. To enable DC braking, you need to set bit 9 in *Bus/Brake Opts* (parameter 13).	Parameter number	80
		File:group	Application:DC Braking/Hold
		Parameter type	destination
		Display	x.x seconds
		Factory default	10.0 seconds
		Minimum value	0.0 seconds
		Maximum value	6553.5 seconds
		Conversion	10=1.0 seconds

1　*DC Brake Time* was added in Version 2.xx.

81	**Motor Speed**		
	Motor Speed contains a filtered version of speed feedback. The value displayed in *Motor Speed* is passed through a low pass filter having a 125 millisecond time constant.	Parameter number	81
		File:group	Monitor:Motor Status
		Parameter type	source
		Display	±x.x rpm
		Factory default	not applicable
		Minimum value	-8 x base motor speed
		Maximum value	+8 x base motor speed
		Conversion	4096 = base motor speed

83 Motor Current

Use *Motor Current* to view the actual RMS value of the motor current as determined from the LEM current sensors. This data is averaged and updated every 50 milliseconds.

Parameter number	83
File:group	Monitor:Motor Status
Parameter type	source
Display	x.x amp
Factory default	not applicable
Minimum value	0.0 amps
Maximum value	6553.5 amps
Conversion	4096 = rated inverter amps

84 DC Bus Voltage

DC Bus Voltage represents the actual bus voltage in volts as read by the software from an analog input port.

Parameter number	84
File:group	Monitor:Drive/Inv Status
Parameter type	source
Display	x volts
Factory default	not applicable
Minimum value	0 volts
Maximum value	1000 volts
Conversion	1 = 1

85 Motor Voltage

Use *Motor Voltage* to view the actual line-to-line fundamental RMS value of motor voltage. This data is averaged and updated every 50 milliseconds.

Parameter number	85
File:group	Monitor:Motor Status
Parameter type	source
Display	x volt
Factory default	not applicable
Minimum value	0 volts
Maximum value	+3000 volts
Conversion	1 = 1

86 Motor Torque %

Use *Motor Torque %* to view the calculated value of motor torque as determined by the drive. The actual value of the motor torque is within 5% of this value. This data is updated every 2 milliseconds.

Parameter number	86
File:group	Monitor:Motor Status
Parameter type	source
Display	±x.x% trq
Factory default	not applicable
Minimum value	-800.0%
Maximum value	+800.0%
Conversion	4096 = 100.0%

88 Motor Flux %

Use *Motor Flux %* to view the level of motor field flux calculated by the drive.

Parameter number	88
File:group	Monitor:Motor Status
Parameter type	source
Display	x.x%
Factory default	not applicable
Minimum value	12.5%
Maximum value	100.0%
Conversion	4096 = 100.0%

89 Motor Frequency

Use *Motor Frequency* to view the actual value of motor stator frequency in Hz.

Parameter number	89
File:group	Monitor:Motor Status
Parameter type	source
Display	x.xxx Hz
Factory default	not applicable
Minimum value	-250.000 Hz
Maximum value	+250.000 Hz
Conversion	128 = 1.000

90 Motor Power %

Motor Power % is the calculated product of torque reference times motor speed feedback. A 125 millisecond filter is applied to this result. Positive values indicate motoring power; negative values indicate regenerative power.

Parameter number	90
File:group	Monitor:Motor Status
Parameter type	source
Display	±x.x% PWR
Factory default	not applicable
Minimum value	-800.0%
Maximum value	+800.0%
Conversion	4096 = 100.0%

91 Iq %

Iq % shows the value of torque current reference that is present at the output of the current rate limiter. 100% is equal to 1 per unit (pu) rated motor torque.

Parameter number	91
File:group	none
Parameter type	source
Display	±x.x%
Factory default	not applicable
Minimum value	-800.0%
Maximum value	+800.0%
Conversion	4096 = 100.0%

附录 A 1336 Impact 变频器常用参数

96 An In 1 Value
Use *An In 1 Value* to view the converted analog value of the input at analog input 1.

Parameter number	96
File:group	Interface/Comm:Analog Inputs
Parameter type	source
Display	±x
Factory default	not applicable
Minimum value	-32767
Maximum value	+32767
Conversion	1 = 1

97 An In 1 Offset
Use *An In 1 Offset* to set the offset applied to the raw analog value of the analog input 1 before the scale factor is applied. This lets you shift the range of the analog input.

Parameter number	97
File:group	Interface/Comm:Analog Inputs
Parameter type	linkable destination
Display	±x.xxx volts
Factory default	0.000 volts
Minimum value	-19.980 volts
Maximum value	+19.980 volts
Conversion	205 = 1.000

98 An In 1 Scale
Use *An In 1 Scale* to set the scale factor or gain for analog input 1. The value of analog input 1 is converted to +2048 and then the scale is applied. This provides an effective digital range of ±32767.

Parameter number	98
File:group	Interface/Comm:Analog Inputs
Parameter type	linkable destination
Display	±x.xxx
Factory default	+2.000
Minimum value	-16.000
Maximum value	+16.000
Conversion	2048 = 1.000

99 An In 2 Value
Use *An In 2 Value* to view the converted analog value of the input at analog input 2.

Parameter number	99
File:group	Interface/Comm:Analog Inputs
Parameter type	source
Display	±x
Factory default	not applicable
Minimum value	-32767
Maximum value	+32767
Conversion	1 = 1

100 An In 2 Offset
Use *An In 2 Offset* to set the offset applied to the raw analog value of analog input 2 before the scale factor is applied. This lets you shift the range of the analog input.

Parameter number	100
File:group	Interface/Comm:Analog Inputs
Parameter type	linkable destination
Display	±x.xxx volts
Factory default	0.000 volts
Minimum value	-19.980 volts
Maximum value	+19.980 volts
Conversion	205 = 1.000

101 An In 2 Scale
Use *An In 2 Scale* to set the scale factor or gain for analog input 2. The value of analog input 2 is converted to +2048 and then the scale is applied. This provides an effective digital range of ±32767.

Parameter number	101
File:group	Interface/Comm:Analog Inputs
Parameter type	linkable destination
Display	±x.xxx
Factory default	+2.000
Minimum value	-16.000
Maximum value	+16.000
Conversion	2048 = 1.000

105 An Out 1 Value
Use *An Out 1 Value* to convert a +32767 digital value to a +10 volt output. This is the value of the analog output number 1.

Parameter number	105
File:group	Interface/Comm:Analog Outputs
Parameter type	linkable destination
Display	±x
Factory default	+0
Minimum value	-32767
Maximum value	+32767
Conversion	1 = 1

106 An Out 1 Offset
Use *An Out 1 Offset* to set the offset applied to the raw analog output 1. The offset is applied after the scale factor.

Parameter number	106
File:group	Interface/Comm:Analog Outputs
Parameter type	linkable destination
Display	±x.xxx volts
Factory default	+0.000 volts
Minimum value	-20.000 volts
Maximum value	+20.000 volts
Conversion	205 = 1.000

107	**An Out 1 Scale**	Parameter number	107
	Use *An Out 1 Scale* to set the scale factor or gain for analog output 1. A +32767 digital value is converted by the scale factor. This allows an effective digital range of +2048 which is then offset to provide a +10 volt range.	File:group	Interface/Comm:Analog Outputs
		Parameter type	linkable destination
		Display	±x.xxx
		Factory default	+0.500
		Minimum value	-1.000
		Maximum value	+1.000
		Conversion	32767 = 1.000

108	**An Out 2 Value**	Parameter number	108
	Use *An Out 2 Value* to convert a +32767 digital value to a +10 volt output. This is the value of the analog output number 2.	File:group	Interface/Comm:Analog Outputs
		Parameter type	linkable destination
		Display	±x
		Factory default	+0
		Minimum value	-32767
		Maximum value	+32767
		Conversion	1 = 1

109	**An Out 2 Offset**	Parameter number	109
	Use *An Out 2 Offset* to set the offset applied to the raw analog output 2. The offset is applied after the scale factor.	File:group	Interface/Comm:Analog Outputs
		Parameter type	linkable destination
		Display	±x.xxx volts
		Factory default	+0.000 volts
		Minimum value	-19.980 volts
		Maximum value	+19.980 volts
		Conversion	205 = 1.000

110	**An Out 2 Scale**	Parameter number	110
	Use *An Out 2 Scale* to set the scale factor or gain for analog output 2. A +32767 digital value is converted by the scale factor. This allows an effective digital range of +2048 which is then offset to provide a +10 volt range.	File:group	Interface/Comm:Analog Outputs
		Parameter type	linkable destination
		Display	±x.xxx
		Factory default	+0.500
		Minimum value	-1.000
		Maximum value	+1.000
		Conversion	32767 = 1.000

111	**mA Out Value**	Parameter number	111
	Use *mA Out Value* to convert a +32767 digital value to a 4 – 20 mA output. This is the value of the mA output.	File:group	Interface/Comm:Analog Outputs
		Parameter type	linkable destination
		Display	±x
		Factory default	+0
		Minimum value	-32767
		Maximum value	+32767
		Conversion	1 = 1

112	**mA Out Offset**	Parameter number	112
	Use *mA Out Offset* to set the offset applied to the raw milli amp output. The offset is applied after the scale factor.	File:group	Interface/Comm:Analog Outputs
		Parameter type	linkable destination
		Display	±x.xxx mA
		Factory default	+0.000 mA
		Minimum value	-32.000 mA
		Maximum value	+32.000 mA
		Conversion	128 = 1.000

113	**mA Out Scale**	Parameter number	113
	Use *mA Out Scale* to set the scale factor or gain for milli amp output. A +32767 digital value is converted by the scale factor. This allows an effective digital range of +2048 which is then offset to provide a +20 mA range.	File:group	Interface/Comm:Analog Outputs
		Parameter type	linkable destination
		Display	±x.xxx
		Factory default	+0.500
		Minimum value	-1.000
		Maximum value	+1.000
		Conversion	32767 = 1.000

115	**Relay Setpoint 1**	Parameter number	115
	Relay Setpoint 1 lets you specify the setpoint threshold for either speed or current. *Relay Setpoint 1* is only active if *Relay Config 1* (parameter 114) is set to a value of 25, 26, 27, or 28.	File:group	Interface/Comm:Digital Config
		Parameter type	linkable destination
		Display	±x.x%
		Factory default	+0.0%
		Minimum value	-800.0%
		Maximum value	+800.0%
		Conversion	4096 = 100.0%

116	**L Option Mode**				Parameter number			116

Use *L Option Mode* to select the functions of L Option inputs at TB3. If you change the value of *L Option Mode*, you must cycle power before the change will take effect.

The following is the mode information:

					Parameter number			116
					File:group		Interface/Comm:Digital Config	
					Parameter type			destination
					Display			x
					Factory default			1
					Minimum value			1
					Maximum value			32
					Conversion			1 = 1

Mode	TB3-19	TB3-20	TB3-22	TB3-23	TB3-24	TB3-26	TB3-27	TB3-28
1	Status	Stop	Status	Status	Status	Status	Status	Status
2	Start	Stop	Rev/Fwd	Jog	Ext Fault	Spd 3	Spd 2	Spd 1
3	Start	Stop	Rev/Fwd	2/1Stop Type	Ext Fault	Spd 3	Spd 2	Spd 1
4	Start	Stop	Rev/Fwd	2/1 Accel	Ext Fault	2/1 Decel	Spd 2	Spd 1
5	Start	Stop	Rev/Fwd	Pot Up	Ext Fault	Pot Dn	Spd 2	Spd 1
6	Start	Stop	Rev/Fwd	Jog	Ext Fault	Loc/Rem	Spd 2	Spd 1
7	Start	Stop	Rev	Fwd	Ext Fault	Jog	Spd 2	Spd 1
8	Start	Stop	Rev	Fwd	Ext Fault	Spd 3	Spd 2	Spd 1
9	Start	Stop	Pot Up	Pot Dn	Ext Fault	Spd 3	Spd 2	Spd 1
10	Start	Stop	Rev	Fwd	Ext Fault	Pot Up	Pot Dn	Spd 1
11	Start	Stop	1st Accel	2nd Accel	Ext Fault	1st Decel	2nd Decel	Spd 1
12	Run Fwd	Stop	Run Rev	Loc/Rem	Ext Fault	Spd 3	Spd 2	Spd 1
13	Run Fwd	Stop	Run Rev	2/1 Stop Type	Ext Fault	Spd 3	Spd 2	Spd 1
14	Run Fwd	Stop	Run Rev	2/1 Accel	Ext Fault	2/1 Decel	Spd 2	Spd 1
15	Run Fwd	Stop	Run Rev	Pot Up	Ext Fault	Pot Dn	Spd 2	Spd 1
16	Run Fwd	Stop	Run Rev	Loc/Rem	Ext Fault	2/1 Stop Type	Spd 2	Spd 1
17	Start	Stop	Rev/Fwd	Proc Trim	Ext Fault	Ramp	Spd 2	Spd 1
18	Start	Stop	Rev/Fwd	Flux En	Ext Fault	Reset	Spd 2	Spd 1
19	Start	Stop	Spd/Trq 3	Spd/Trq 2	Ext Fault	Spd/Trq 1	Proc Trim	Spd 1
20	Start	Stop	Spd/Trq 3	Spd/Trq 2	Ext Fault	Spd/Trq 1	Flux En	Spd 1
21	Rev		Ramp	Fwd	Ext Fault	Ramp	Reset	Spd 1
22	Start	Stop	Spd/Trq 3	Spd/Trq 2	Ext Fault	Spd/Trq 1	Spd 2	Spd 1
23	Run Fwd	Stop	Run Rev	Proc Trim	Ext Fault	Reset	Spd 2	Spd 1
24	Run Fwd	Stop	Run Rev	Flux En	Ext Fault	Reset	Spd 2	Spd 1
25	Run Fwd	Stop	Run Rev	Proc Trim	Ext Fault	Ramp	Spd 2	Spd 1
26[1]	Run Fwd	Stop	Run Rev	Jog	Ext Fault	Spd 3	Spd 2	Spd 1
27[2]	Start	Stop	Rev/Fwd	Pot Up	Ext Fault	Pot Dn	Spd 2	Spd 1
28[2]	Start	Stop	Pot Up	Pot Dn	Ext Fault	Spd 3	Spd 2	Spd 1
29[2]	Rev		Ramp	Fwd	Ext Fault	Pot Up	Pot Dn	Spd 1
30[2]	Run Fwd	Stop	Run Rev	Pot Up	Ext Fault	Pot Dn	Spd 2	Spd 1
31[3]	Step Trigger	Not Stop	Step Trigger	Step Trigger	Not Ext Flt	Step Trigger	Step Trigger	Step Trigger
32[3]	Start		Not Stop	Step Trigger	Step Trigger	Not Ext Flt	Profile Enable	Run Sequence	Step Hold

1 Added for Version 2.01.
2 Added for Version 2.02.
3 Added for Version 4.01.

117	**L Option In Sts**	Parameter number	117

Use *L Option In Sts* to view the status of the L Option inputs.

	File:group	Interface/Comm:Digital Config
	Parameter type	source
	Display	bits
	Factory default	not applicable
	Minimum value	00000000.00000000
	Maximum value	00000001.11111111
	Conversion	1 = 1

Bit	Description	Bit	Description	Bit	Description	Bit	Description
0	TB3-19	3	TB3-23	6	TB3-27	9 – 15	Reserved
1	TB3-20	4	TB3-24	7	TB3-28		Leave 0.
2	TB3-22	5	TB3-26	8	TB3-30 (enable)		

118	**Mop Increment**	Parameter number	118

Use *Mop Increment* to set the rate of increase or decrease to the Manually Operated Potentiometer (MOP) value based on rpm/second. *Mop Increment* is only used when the value of *L Option Mode* (parameter 116) is 5, 9, 10, or 15.

	File:group	Interface/Comm:Digital Config
	Parameter type	linkable destination
	Display	x.x rpm (rpm/second)
	Factory default	10% of base motor speed
	Minimum value	0.0
	Maximum value	base motor speed
	Conversion	4096 = base motor speed

119	**Mop Value**	Parameter number	119

Use *Mop Value* to view the Manually Operated Potentiometer (MOP) value. You need to link *Mop Value* to a reference, such as *Speed Ref 1* (parameter 29) for the drive to follow the Mop command for speed.

	File:group	Interface/Comm:Digital Config
	Parameter type	source
	Display	±x.x rpm
	Factory default	not applicable
	Minimum value	0.0
	Maximum value	base motor speed
	Conversion	4096 = base motor speed

133	**SP An In1 Select**	Parameter number	133
	Use *SP An In1 Select* to select which SCANport analog device is used in *SP An In1 Value* (parameter 134).	File:group	Interface/Comm:SCANport Analog
		Parameter type	linkable destination
		Display	x
		Factory default	1
		Minimum value	1
		Maximum value	6
		Conversion	1 = 1

Value	Description	Value	Description	Value	Description
1	SP 1 Use SCANport device 1.	3	SP 3 Use SCANport device 3.	5	SP 5 Use SCANport device 5.
2	SP 2 Use SCANport device 2.	4	SP 4 Use SCANport device 4.	6	SP 6 Use SCANport device 6.

134	**SP An In1 Value**	Parameter number	134
	Use *SP An In1 Value* to view the analog value of the SCANport device selected in *SP An In1 Select* (parameter 133). You need to link *SP An In1 Value* to a parameter such as *Speed Ref 1* (parameter 29).	File:group	Interface/Comm:SCANport Analog
		Parameter type	source
		Display	±x
		Factory default	not applicable
		Minimum value	-32767
		Maximum value	+32767
		Conversion	1 = 1

135	**SP An In1 Scale**	Parameter number	135
	Use *SP An In1 Scale* to scale *SP An In1 Value* (parameter 134).	File:group	Interface/Comm:SCANport Analog
		Parameter type	linkable destination
		Display	±x.xxx
		Factory default	+0.125
		Minimum value	-1.000
		Maximum value	+1.000
		Conversion	32767 = 1.000

136	**SP An In2 Select**	Parameter number	136
	Use *SP An In2 Select* to select which SCANport analog device is used in *SP An In2 Value* (parameter 137).	File:group	Interface/Comm:SCANport Analog
		Parameter type	linkable destination
		Display	x
		Factory default	6
		Minimum value	1
		Maximum value	6
		Conversion	1 = 1

Value	Description	Value	Description	Value	Description
1	SP 1 Use SCANport device 1.	3	SP 3 Use SCANport device 3.	5	SP 5 Use SCANport device 5.
2	SP 2 Use SCANport device 2.	4	SP 4 Use SCANport device 4.	6	SP 6 Use SCANport device 6.

137	**SP An In2 Value**	Parameter number	137
	Use *SP An In2 Value* to view the analog value of the SCANport device selected in *SP An In2 Select* (parameter 136). You need to link *SP An In2 Value* to a parameter such as *Speed Ref 1* (parameter 29).	File:group	Interface/Comm:SCANport Analog
		Parameter type	source
		Display	±x
		Factory default	not applicable
		Minimum value	-32767
		Maximum value	+32767
		Conversion	1 = 1

138	**SP An In2 Scale**	Parameter number	138
	Use *SP An In2 Scale* to scale *SP An In2 Value* (parameter 137).	File:group	Interface/Comm:SCANport Analog
		Parameter type	linkable destination
		Display	±x.xxx
		Factory default	+0.125
		Minimum value	-1.000
		Maximum value	+1.000
		Conversion	32767 = 1.000

139	**SP An Output**	Parameter number	139
	Use *SP An Output* to view the analog value that is sent to all SCANport devices. **Note:** If a link is made or changed, you may have to power cycle the SCANport terminals to display the correct information.	File:group	Interface/Comm:SCANport Analog
		Parameter type	linkable destination
		Display	±x
		Factory default	+0
		Minimum value	-32767
		Maximum value	+32767
		Conversion	1 = 1

附录 A 1336 Impact 变频器常用参数

156 Autotune Status

Autotune Status provides information about the auto-tune procedure.

Parameter number	156
File:group	Autotune/Autotune Status
Parameter type	source
Display	bits
Factory default	not applicable
Minimum value	00000000.00000000
Maximum value	00110000.11111111
Conversion	1 = 1

The bits are defined as follows:

Bit	Description	Bit	Description	Bit	Description
0	**Executing** — Auto-tune is currently executing.	4	**Flux Active** — The motor has flux.	8 – 11	**Reserved** — Leave 0.
1	**Complete** — Auto-tune has completed.	5	**Not Ready** — The drive is not ready to start auto-tune.	12	**Timeout** — Auto-tune timed out. The inertia test failed to accelerate the load.
2	**Fail** — An error was encountered.	6	**Not Zero Spd** — The drive cannot start auto-tune.	13	**No Trq Lim** — The inertia test failed to reach the torque limit.
3	**Abort** — Auto-tune was aborted by a stop command.	7	**Running** — The motor is running.	14 – 15	**Reserved** — Leave 0.

158 Ki Speed Loop

Use *Ki Speed Loop* to control the integral error gain of the speed regulator.

The 1336 IMPACT drive automatically adjusts *Ki Speed Loop* when you enter a non-zero value for *Spd Desired BW* (parameter 161). Normally, you should adjust *Spd Desired BW* and let the drive calculate the gains. If manual adjustment is needed (for example, if the inertia cannot be determined), the drive sets *Spd Desired BW* to zero for you when this gain is changed.

Parameter number	158
File:group	Control:Speed Regulator
Parameter type	linkable destination
Display	x.x
Factory default	8.0
Minimum value	0.0
Maximum value	4095.9
Conversion	8 = 1.0

159 Kp Speed Loop

Use *Kp Speed Loop* to control the proportional error gain of the speed regulator.

The 1336 IMPACT drive automatically adjusts *Kp Speed Loop* when you enter a non-zero value for *Spd Desired BW* (parameter 161). Normally, you should adjust *Spd Desired BW* and let the drive calculate the gains. If manual gain adjustment is needed (for example, if the inertia cannot be determined), the drive sets *Spd Desired BW* to zero for you when this gain is changed.

Parameter number	159
File:group	Control:Speed Regulator
Parameter type	linkable destination
Display	x.x
Factory default	8.0
Minimum value	0.0
Maximum value	200.0
Conversion	8 = 1.0

160 Kf Speed Loop

Use *Kf Speed Loop* to control the feed forward gain of the speed regulator. Setting the Kf gain to less than one reduces speed feedback overshoot in response to a step change in speed reference.

Parameter number	160
File:group	Control:Speed Regulator
Parameter type	linkable destination
Display	x.xxx
Factory default	1.000
Minimum value	0.500
Maximum value	1.000
Conversion	65535 = 1.0

161 Spd Desired BW

Use *Spd Desired BW* to specify the speed loop bandwidth and to determine the dynamic behavior of the speed loop. As you increase the bandwidth, the speed loop becomes more responsive and can track a faster changing speed reference.

As you adjust the bandwidth setting, the 1336 IMPACT drive calculates and changes *Ki Speed Loop* (parameter 158) and *Kp Speed Loop* (parameter 159) gains. A zero bandwidth setting lets you adjust the speed loop gains independent of bandwidth for custom tuning applications.

Note: You must have the correct *Total Inertia* (parameter 157) entered before adjusting the speed loop bandwidth. *Total Inertia* is measured by the autotune (startup) routine.

Parameter number	161
File:group	Control:Speed Regulator / Autotune:Autotune Results
Parameter type	linkable destination
Display	x.xx radians/second
Factory default	5.00 radians/second
Minimum value	0.00 radians/second
Maximum value	calculated
Conversion	100 = 1

164 Autotune Torque

Use *Autotune Torque* to specify the motor torque that is applied to the motor during the flux current and inertia tests.

Parameter number	164
File:group	Autotune:Autotune Setup
Parameter type	destination
Display	x.x%
Factory default	50.0%
Minimum value	25.0%
Maximum value	100.0%
Conversion	4096 = 100.0%

165 Autotune Speed

Use *Autotune Speed* to set the maximum speed of the motor during the flux current and inertia tests.

Parameter number	165
File:group	Autotune/Autotune Setup
Parameter type	destination
Display	±x.x rpm
Factory default	base motor speed x 0.85
Minimum value	base motor speed x 0.3
Maximum value	base motor speed
Conversion	4096 = base motor speed

166 Stator Resistance

Enter the sum of the stator and cable resistances of the motor in per unit (percent representation). The auto-tune procedure measures the stator resistance during the quick motor tune portion of start up.

Parameter number	166
File:group	Motor/Inverter:Motor Constants
	Autotune/Autotune Results
Parameter type	destination
Display	x.xx%
Factory default	1.49%
Minimum value	0.00%
Maximum value	100.00%
Conversion	4096 = 100.00%

167 Leak Inductance

Enter the sum of the motor stator and rotor leakage inductances and the motor cable inductance in per unit (percent representation). The auto-tune procedure measures the leakage inductance during the quick motor tune portion of start up.

Parameter number	167
File:group	Motor/Inverter:Motor Constants
	Autotune:Autotune Results
Parameter type	destination
Display	x.xx%
Factory default	17.99%
Minimum value	0.00%
Maximum value	100.00%
Conversion	4096 = 100.00%

168 Flux Current

Use *Flux Current* to specify the magnetizing current that produces rated flux in the motor in a per unit (percent representation). The auto-tune procedure measures the flux current during the quick motor tune portion of start up.

Parameter number	168
File:group	Motor/Inverter:Motor Constants
	Autotune:Autotune Results
Parameter type	destination
Display	x.xx%
Factory default	30.00%
Minimum value	0.00%
Maximum value	75.00%
Conversion	4096 = 100.00%

169 Slip Gain

Use *Slip Gain* to fine tune the slip constant of the motor to improve speed regulation in encoderless mode.

Parameter number	169
File:group	Motor/Inverter:Motor Constants
	Autotune:Autotune Results
Parameter type	destination
Display	x.x%
Factory default	100.0%
Minimum value	0.0%
Maximum value	400.0%
Conversion	1024 = 100.0%

170 Vd Max

Use *Vd Max* to view the maximum D axis voltage allowed on the motor. The auto-tune routine calculates the value of *Vd Max*. You should not change this value.

Vd is short for flux axis voltage.

Parameter number	170
File:group	none
Parameter type	destination
Display	x.x volts
Factory default	calculated
Minimum value	0.0 volts
Maximum value	468.8 volts
Conversion	16 = 1.0

171 Vq Max

Use *Vq Max* to view the Q axis voltage at which the motor enters field weakening. The auto-tune routine calculates the value of *Vq Max*. You should not change this value.

Vq is short for torque axis voltage.

Parameter number	171
File:group	none
Parameter type	destination
Display	x.x volts
Factory default	calculated
Minimum value	0.0 volts
Maximum value	468.8 volts
Conversion	16 = 1.0

附录 A 1336 Impact 变频器常用参数

177 Ki Freq Reg
Ki Freq Reg contains the integral gain of the frequency regulator in encoderless mode. Do not change the value of this parameter.

Parameter number	177
File:group	none
Parameter type	destination
Display	x
Factory default	300
Minimum value	0
Maximum value	32767
Conversion	1 = 1

178 Kp Freq Reg
Kp Freq Reg contains the proportional gain of the frequency regulator in encoderless mode. Do not change the value of this parameter.

Parameter number	178
File:group	none
Parameter type	destination
Display	x
Factory default	800
Minimum value	0
Maximum value	32767
Conversion	1 = 1

179 Kf Freq Reg
Kf Freq Reg contains the feed-forward gain of the frequency regulator in encoderless mode. Do not change the value of this parameter.

Parameter number	179
File:group	none
Parameter type	destination
Display	x.x
Factory default	1.0
Minimum value	0.0
Maximum value	128.0
Conversion	256 = 1.0

182 An In1 Filter BW[1]
Use *An In1 Filter BW* to use a low pass filter on the analog input 1. This filter adjusts the bandwidth to get better filtering. By using the low pass filter, you lose some bandwidth, but the value becomes more stable.

[1] *An In1 Filter BW* was added in Version 2.xx.

Parameter number	182
File:group	Interface/Comm:Analog Inputs
Parameter type	linkable destination
Display	x.x radians per second
Factory default	0.0 radians per second
Minimum value	0.0 radians per second
Maximum value	200.0 radians per second
Conversion	10 = 1

183 An In2 Filter BW[1]
Use *An In2 Filter BW* to use a low pass filter on the analog input 2. This filter adjusts the bandwidth to get better filtering. By using the low pass filter, you lose some bandwidth, but the value becomes more stable.

[1] *An In1 Filter BW* was added in Version 2.xx.

Parameter number	183
File:group	Interface/Comm:Analog Inputs
Parameter type	linkable destination
Display	x.x radians per second
Factory default	0.0 radians per second
Minimum value	0.0 radians per second
Maximum value	200.0 radians per second
Conversion	10 = 1

184 mA In Filter BW[1]
Use *mA In Filter BW* to use a low pass filter on the 4 – 20 mA input. This filter adjusts the bandwidth to get better filtering. By using the low pass filter, you lose some bandwidth, but the value becomes more stable.

[1] *mA In Filter BW* was added in Version 2.xx.

Parameter number	184
File:group	Interface/Comm:Analog Inputs
Parameter type	linkable destination
Display	x.x radians per second
Factory default	0.0 radians per second
Minimum value	0.0 radians per second
Maximum value	200.0 radians per second
Conversion	10 = 1

193 Start Dwell Spd[1]
Start Dwell Spd lets you set the speed that the drive immediately outputs when a start command is issued. No acceleration ramp is used. You must enter a time value in *Start Dwell Time* (parameter 194).

[1] *Start Dwell Spd* was added in Version 2.xx.

Parameter number	193
File:group	Control:Drive Logic Sel
Parameter type	linkable destination
Display	±x.x rpm
Factory default	+0.0 rpm
Minimum value	-0.1 x base motor speed
Maximum value	+0.1 x base motor speed
Conversion	4096 = base motor speed

194 Start Dwell Time[1]
Start Dwell Time lets you specify how long you want the drive to continue using *Start Dwell Spd* (parameter 193) before ramping to whichever speed reference you have selected (speed references 1 through 7).

[1] *Start Dwell Time* was added in Version 2.xx.

Parameter number	194
File:group	Control:Drive Logic Sel
Parameter type	linkable destination
Display	x.x seconds
Factory default	0.0 seconds
Minimum value	0.0 seconds
Maximum value	10.0 seconds
Conversion	seconds x 10

198 Function In1[1]

Use *Function In1* to provide input into the function block that is provided with the 1336 IMPACT drive. You can choose to either evaluate the input value or pass the value directly to the function block.

To evaluate *Function In1*, you need to also use *Func 1 Mask/Val* (parameter 199) and *Func 1 Eval Sel* (parameter 200).

To pass the value directly to the function block, enter a value of 0 into *Func 1 Eval Sel*.

[1] *Function In1* was added in Version 2.xx.

Parameter number	198
File:group	Application:Prog Function
Parameter type	linkable destination
Conversion	1 = 1
If *Func 1 Eval Sel* (parameter 200) is 0 or 6 – 11, then:	
Display	±x
Factory default	0
Minimum value	-32767
Maximum value	+32767
If *Func 1 Eval Sel* (parameter 200) is 1 – 5, then:	
Display	bits
Factory default	00000000.00000000
Minimum value	00000000.00000000
Maximum value	11111111.11111111
If *Func 1 Eval Sel* (parameter 200) is 12 – 15, then:	
Display	x
Factory default	0
Minimum value	0
Maximum value	65535

199 Func 1 Mask/Val[1]

Use *Func 1 Mask/Val* to enter a mask or value to compare *Function In1* (parameter 198) to, according to the value you select in *Func 1 Eval Sel* (parameter 200).

[1] *Func 1 Mask/Val* was added in Version 2.xx.

Parameter number	199
File:group	Application:Prog Function
Parameter type	linkable destination
Conversion	1 = 1
If *Func 1 Eval Sel* (parameter 200) is 0 or 6 – 11, then:	
Display	±x
Factory default	-1
Minimum value	-32767
Maximum value	+32767
If *Func 1 Eval Sel* (parameter 200) is 1 – 5, then:	
Display	bits
Factory default	11111111.11111111
Minimum value	00000000.00000000
Maximum value	11111111.11111111
If *Func 1 Eval Sel* (parameter 200) is 12 – 15, then:	
Display	x
Factory default	65535
Minimum value	0
Maximum value	65535

200 Func 1 Eval Sel[1]

Func 1 Eval Sel lets you choose how you want to evaluate *Function In1* (parameter 198).

[1] *Func 1 Eval Sel* was added in Version 2.xx.

Parameter number	200
File:group	Application:Prog Function
Parameter type	destination
Display	x
Factory default	0
Minimum value	0
Maximum value	17
Conversion	1 = 1

Value	Description	Value	Description	Value	Description
0	**None** Pass the value directly on to the function block.	6	**I=V** Check to see if *Function In1* is equal to *Func 1 Mask/Val*.	12	**Unsign I<V** Check to see if the unsigned value of *Function In1* is less than the value of *Func 1 Mask/Val*.
1	**Mask** Mask specific bits.	7	**I≠V** Check to see if *Function In1* is not equal to *Func 1 Mask/Val*.	13	**Unsign I≤V** Check to see if the unsigned value of *Function In1* is less than or equal to the value of *Func 1 Mask/Val*.
2	**All Bits On** Check to make sure that all bits that are set (on) in *Func 1 Mask/Val* (parameter 199) are set in *Function In1* (parameter 198).	8	**Signed I<V** Check to see if the signed value of *Function In1* is less than the value of *Func 1 Mask/Val*.	14	**Unsign I>V** Check to see if the unsigned value of *Function In1* is greater than the value of *Func 1 Mask/Val*.
3	**All Bits Off** Check to make sure that all bits that are set in *Func 1 Mask/Val* are clear in *Function In1*.	9	**Signed I≤V** Check to see if the signed value of *Function In1* is less than or equal to the value of *Func 1 Mask/Val*.	15	**Unsign I≥V** Check to see if the unsigned value of *Function In1* is greater than or equal to the value of *Func 1 Mask/Val*.
4	**Any Bit On** Check to make sure that at least one of the bits that are set in *Func 1 Mask/Val* is set in *Function In1*.	10	**Signed I>V** Check to see if the signed value of *Function In1* is greater than the value of *Func 1 Mask/Val*.	16	**Invert** Pass the opposite value on to the function block
5	**Any Bit Off** Check to make sure that at least one of the bits that are set in *Func 1 Mask/Val* is clear in *Function In1*.	11	**Signed I≥V** Check to see if the signed value of *Function In1* is greater than or equal to the value of *Func 1 Mask/Val*.	17	**Absolute** Pass a positive value on to the function block.

附录 A　1336 Impact 变频器常用参数

201　Function In2[1]

Use *Function In2* to provide input into the function block that is provided with the 1336 IMPACT drive. You can choose to either evaluate *Function In2* or pass the value directly to the function block.

To evaluate *Function In2*, you need to also use *Func 2 Mask/Val* (parameter 202) and *Func 2 Eval Sel* (parameter 203).

To pass the value directly to the function block, enter a value of 0 into *Func 2 Eval Sel*.

[1] *Function In2* was added in Version 2.xx.

Parameter number	201
File:group	Application:Prog Function
Parameter type	linkable destination
Conversion	1 = 1
If *Func 2 Eval Sel* (parameter 203) is 0 or 6 – 11, then:	
Display	±x
Factory default	0
Minimum value	-32767
Maximum value	+32767
If *Func 2 Eval Sel* (parameter 203) is 1 – 5, then:	
Display	bits
Factory default	00000000.00000000
Minimum value	00000000.00000000
Maximum value	11111111.11111111
If *Func 2 Eval Sel* (parameter 203) is 12 – 15, then:	
Display	x
Factory default	0
Minimum value	0
Maximum value	65535

204　Function In3[1]

Use *Function In3* to provide input into the function block that is provided with the 1336 IMPACT drive. You can choose to either evaluate the input value or pass the value directly to the function block.

To evaluate *Function In3*, you need to also use *Func 3 Mask/Val* (parameter 205) and *Func 3 Eval Sel* (parameter 206).

To pass the value directly to the function block, enter a value of 0 into *Func 3 Eval Sel*.

[1] *Function In3* was added in Version 2.xx.

Parameter number	204
File:group	Application:Prog Function
Parameter type	linkable destination
Conversion	1 = 1
If *Func 3 Eval Sel* (parameter 206) is 0 or 6 – 11, then:	
Display	±x
Factory default	0
Minimum value	-32767
Maximum value	+32767
If *Func 3 Eval Sel* (parameter 206) is 1 – 5, then:	
Display	bits
Factory default	00000000.00000000
Minimum value	00000000.00000000
Maximum value	11111111.11111111
If *Func 3 Eval Sel* (parameter 206) is 12 – 15, then:	
Display	x
Factory default	0
Minimum value	0
Maximum value	65535

207　Function In4[1]

Use *Function In4* to provide input to the function block that is provided with the 1336 IMPACT drive.

For the timer delay and state machine function blocks, *Function In4* is used to specify how long after the timer off input is received before turning off the timer output. When used for these modes, the timer off signal must be present for as long as you specify in *Function In4*.

For the up/down counter function block, *Function In4* specifies how much to add to the value when *Function In1* (parameter 198) indicates that a rising edge has occurred.

For the multiply/divide function block, *Function In4* specifies whether the function should be performed as a per unit function or as a math function.

For the scale function block, *Function In4* is the upper word of the value that you want to use as either the minimum or maximum value for the output. The lower word of this value is specified in *Function In5* (parameter 208).

[1] *Function In4* was added in Version 2.xx.

Parameter number	207
File:group	Application:Prog Function
Parameter type	linkable destination
Conversion	1 = 1
If *Function Sel* (parameter 212) is 0 – 8, then:	
Display	xxx.xx minutes
Factory default	0.00 minutes
Minimum value	0.00 minutes
Maximum value	655.35 minutes
If *Function Sel* (parameter 212) is 9 – 12, then:	
Display	x
Factory default	0
Minimum value	0
Maximum value	65535
If *Function Sel* (parameter 212) is 13, then:	
Display	±x
Factory default	0
Minimum value	-32767
Maximum value	+32767

Refer to Chapter 10, *Using the Function Block*, for more information.

208　Function In5[1]

Use *Function In5* to provide input to the function block that is provided with the 1336 IMPACT drive.

For the timer delay and state machine function blocks, *Function In5* is used to specify how long after the timer on input is received before turning on the timer output. When used for these modes, the timer on signal must be present for as long as you specify in *Function In5*.

Parameter number	208
File:group	Application:Prog Function
Parameter type	linkable destination
Conversion	1 = 1
If *Function Sel* (parameter 212) is 0 – 8, then:	
Display	xxx.xx minutes
Factory default	0.00 minutes
Minimum value	0.00 minutes
Maximum value	655.35 minutes

For the up/down counter function block, *Function In5* specifies how much to subtract from the value when *Function In2* (parameter 201) indicates that a rising edge has occurred.

For the scale function block, *Function In5* is the lower word of the value that you want to use as either the minimum or maximum value for the output. The upper word of this value is specified in *Function In4* (parameter 207).

1 *Function In5* was added in Version 2.xx.

If *Function Sel* (parameter 212) is 9 – 13, then:	
Display	x
Factory default	0
Minimum value	0
Maximum value	65535

209 Function In6[1]

Use *Function In6* to provide input to the function block that is provided with the the 1336 IMPACT drive.

For the timer delay function block, *Function In6* specifies the value to pass to *Function Output 1* (parameter 213) when the timer delay output is true.

For the state machine function block, *Function In6* is used for the output if the evaluation of *Function In2* (parameter 201) is false and the evaluation of *Function In1* (parameter 198) and the timer on function are true.

For the up/down counter function block, *Function In6* specifies whether the output is a double word (if *Function In6* is true) or a word (if *Function In6* is false).

For the scale function block, *Function In6* is the upper word of the value that you want to use as either the minimum or maximum value for the output. The lower word of this value is specified in *Function In7* (parameter 210).

1 *Function In6* was added in Version 2.xx.

Parameter number	209
File:group	Application:Prog Function
Parameter type	linkable destination
Conversion	1 = 1

If *Function Sel* (parameter 212) is 0 – 10 or 12, then:	
Display	bits
Factory default	00000000.00000000
Minimum value	00000000.00000000
Maximum value	11111111.11111111

If *Function Sel* (parameter 212) is 11, then:	
Display	x
Factory default	0
Minimum value	0
Maximum value	65535

If *Function Sel* (parameter 212) is 13, then:	
Display	±x
Factory default	0
Minimum value	-32767
Maximum value	+32767

212 Function Sel[1]

Use *Function Sel* to select which function you would like the function block to perform.

1 *Function Sel* was added in Version 2.xx.

Parameter number	212
File:group	Application:Prog Function
Parameter type	destination
Display	x
Factory default	0
Minimum value	0
Maximum value	27
Conversion	1 = 1

Value	Description	Value	Description	Value	Description
0	**Or Tmr** Take the OR of input 1 and input 2 and use the result for the timer input.	10	**Max/Min** Compare input 1 with input 2 and based on input 3, output whichever value is larger or smaller.	20	**Or And Add** Take the result of input 1 OR'ed with input 2 and AND with input 3. Then, use the result for the add/sub input.
1	**Nor Tmr** Take the NOR of input 1 and input 2 and use the result for the timer input.	11	**Counter** Count up (input 1) or down (input 2).	21	**And Or Add** Take the result of input 1 AND input 2 and OR with input 3. Then, use the result for the add/sub input.
2	**And Tmr** Take the AND of input 1 and input 2 and use the result for the timer input.	12	**Mult/Div** Multiply input 1 and input 2 and then divide by input 3.	22	**Or Mult** Take the OR of input 1 and input 2 and use the result for the mult/div input.
3	**Nand Tmr** Take the NAND of input 1 and input 2 and use the result for the timer input.	13	**Scale** Scale the value of input 1 from one range to another.	23	**Nor Mult** Take the NOR of input 1 and input 2 and use the result for the mult/div input.
4	**Or And Tmr** Take the result of input 1 OR'ed with input 2 and AND with input 3. Then, use the result for the timer input.	14	**Hysteresis** Create Hysteresis band (In4-Hi, In5-Lo) for Input 1.	24	**And Mult** Take the AND of input 1 and input 2 and use the result for the mult/div input.
5	**And Or Tmr** Take the result of input 1 AND input 2 and OR with input 3. Then, use the result for the timer input.	15	**Band** Create Band (In4-Hi, In5-Lo) for Input 1.	25	**Nand Mult** Take the NAND of input 1 and input 2 and use the result for the mult/div input.
6	**Tmr Or And** Use input 1 for the timer input and OR with input 2. Then, AND with input 3.	16	**Or Add** Take the OR of input 1 and input 2 and use the result for the add/sub input.	26	**Or And Mult** Take the result of input 1 OR'ed with input 2 and AND with input 3. Then, use the result for the mult/div input.
7	**Tmr And Or** Use input 1 for the timer input and AND with input 2. Then, OR with input 3.	17	**Nor Add** Take the NOR of input 1 and input 2 and use the result for the add/sub input.	27	**And Or Mult** Take the result of input 1 AND input 2 and OR with input 3. Then, use the result for the mult/div input.
8	**StateMachine** Change the output value based on the value of input 1/timer and input 2.	18	**And Add** Take the AND of input 1 and input 2 and use the result for the add/sub input.		
9	**Add/Sub** Add input 1 and input 2.	19	**Nand Add** Take the NAND of input 1 and input 2 and use the result for the add/sub input.		

213 Function Output 1[1]

Use *Function Output 1* to view the results of the function block. *Function Output 1* is either a word value or the upper byte of a double word, depending on the value of *Function Sel* (parameter 212).

1 *Function Output 1* was added in Version 2.xx.

Parameter number	213
File:group	Application:Prog Function
Parameter type	source
Factory default	not applicable
Conversion	1 = 1
If *Function Sel* (parameter 212) is 0 – 8, then:	
Display	bits
Minimum value	00000000.00000000
Maximum value	11111111.11111111
If *Function Sel* (parameter 212) is 9, 10, 12 or 13, then:	
Display	±x
Minimum value	-32767
Maximum value	+32767
If *Function Sel* (parameter 212) is 11, then:	
Display	x
Minimum value	0
Maximum value	65535

214 Function Output 2[1]

Use *Function Output 2* to view the results of the function block. *Function Output 2* is the lower byte of a double word *Function Sel* (parameter 212) is 11, 12, or 13.

1 *Function Output 2* was added in Version 2.xx.

Parameter number	214
File:group	Application:Prog Function
Parameter type	source
Display	x
Factory default	not applicable
Minimum value	0
Maximum value	65535
Conversion	1 = 1

附录 B
PowerFlex 40 变频器常用参数

显示组

d001[输出频率]　　　　　　　　　　　　　　　相关参数:d002,d010,P034,P035,P038
T1,T2 和 T3(U,V 和 W)端的输出频率。

参数值	缺省值	只读
	最小值/最大值	0.0/P035[最大频率]
	显示单位	0.1 Hz

d002[命令频率]　　　　　　　　　　　　　　　相关参数:d001,d013,P034,P035,P038
激活频率命令的数值。即使变频器不运行也会显示命令频率。
重要事项:频率命令有许多来源。详情参阅启动和速度基准值控制。

参数值	缺省值	只读
	最小值/最大值	0.0/P035[最大频率]
	显示单位	0.1 Hz

d003[输出电流]
T1,T2 和 T3(U,V 和 W)端的输出电流。

参数值	缺省值	只读
	最小值/最大值	0.00/(变频器额定电流×2)
	显示单位	0.01 A

d004[输出电压]　　　　　　　　　　　　　　　　　　　相关参数:P031,A084,A088
T1,T2 和 T3(U,V 和 W)端的输出电压。

参数值	缺省值	只读
	最小值/最大值	0/变频器额定电压
	显示单位	1 V AC

附录 B　PowerFlex 40 变频器常用参数

d006[变频器状态]　　　　　　　　　　　　　　　　　　　　　　　相关参数：A095

变频器当前的运行状态。

```
  ┌─ 1＝条件真，0＝条件假
  │
  运行            位0
  正向            位1
  加速            位2
  减速            位3
```

参数值	缺省值	只读
	最小值/最大值	0/1
	显示单位	1

d005[直流母线电压]

当前的直流母线电压。

参数值	缺省值	只读
	最小值/最大值	基于变频器额定值
	显示单位	1 V DC

d007[故障代码 1]
d008[故障代码 2]
d009[故障代码 3]

一个代码表明变频器的一种故障，代码会按照故障发生(d007[故障代码 1]＝最近发生的故障)的顺序出现。重复的故障只被记录一次。

参数值	缺省值	只读
	最小值/最大值	F2/F122
	显示单位	F1

d010[过程显示]　　　　　　　　　　　　　　　　　　　　　　　相关参数：d001，A099

 32 位参数

由参数 A009[过程因素]标定输出频率。
输出频率×过程因素＝过程显示

参数值	缺省值	只读
	最小值/最大值	0.00/9 999
	显示单位	0.01～1

d013[控制输入状态]　　　　　　　　　　　　　　　　　　　　相关参数：d002，P034，P035

变频器端子排输入的状态。

重要事项：实际的控制命令可能来自某个源，而不是控制端子块。

```
0000
     │  1=有输入，0=无输入
     ├─ 起动/正向运行输入(I/O端子02)      位0
     ├─ 方向/反向运行输入(I/O端子03)      位1
     ├─ 停止输入(1) (I/O端子01)           位2
     └─ 动态制动晶闸管工作                位3
```

为了启动变频器，停止输入必须存在。
当此位为1时，变频器可以启动。
当此位为0时，变频器将停止。

参数值	缺省值	只读
	最小值/最大值	0/1
	显示单位	1

d014[数字量输入状态]   相关参数：A051—A054

控制端子排数字量输入端的状态。

重要事项：实际的控制命令可能来自某个源，而不是控制端子块。

```
0000
     │  1=有输入，0=无输入
     ├─ 数字量输入1选择(I/O端子05)       位0
     ├─ 数字量输入2选择(I/O端子06)       位1
     ├─ 数字量输入3选择(I/O端子07)       位2
     └─ 数字量输入4选择(I/O端子08)       位3
```

参数值	缺省值	只读
	最小值/最大值	0/1
	显示单位	1

d015[通信状态]   相关参数：A103—A107

通信端口的状态。

```
0000
     │  1=有输入，0=无输入
     ├─ 数字量输入1选择(I/O端子05)       位0
     ├─ 数字量输入2选择(I/O端子06)       位1
     ├─ 数字量输入3选择(I/O端子07)       位2
     └─ 数字量输入4选择(I/O端子08)       位3
```

参数值	缺省值	只读
	最小值/最大值	0/1
	显示单位	1

附录 B　PowerFlex 40 变频器常用参数

d016[控制板软件版本]

主控板的软件版本。

参数值	缺省值	只读
	最小值/最大值	1.00/99.99
	显示单位	0.01

d017[变频器类型]

供罗克韦尔自动化现场技术服务人员使用。

参数值	缺省值	只读
	最小值/最大值	1 001/9 999
	显示单位	1

d018[累计运转时间]

变频器输出功率的累计时间。时间以 10 h 为增量显示。

参数值	缺省值	只读
	最小值/最大值	0/9 999 h
	显示单位	1＝10 h

d019[测试点数据]　　　　　　　　　　　　　　　　　　　相关参数：A102

在参数 A102[测试点选择]中所选择的功能的当前数值。

参数值	缺省值	只读
	最小值/最大值	0/FFFF
	显示单位	1(16 进制)

d020[模拟量输入 0~10 V]　　　　　　　　　　　　　　　相关参数：A110,A111

I/O 端子 13 的当前电压值(100.0%＝10 V)。

参数值	缺省值	只读
	最小值/最大值	0.0/100.0%
	显示单位	0.10%

d021[模拟量输入 4~20 mA]　　　　　　　　　　　　　　相关参数：A112,A113

I/O 端子 15 的当前电流值(0.0%＝4 mA,100.0%＝20 mA)。

参数值	缺省值	只读
	最小值/最大值	0.0/100.0%
	显示单位	0.10%

d022[输出功率]

T1,T2 和 T3(U,V 和 W)端的输出功率。

参数值	缺省值	只读
	最小值/最大值	0.00/(变频器额定功率×2)
	显示单位	0.01 kW

d023[输出功率因数]

电动机电压和电动机电流之间的电角度。

参数值	缺省值	只读
	最小值/最大值	0.0/180.0 度
	显示单位	0.1 度

d024[变频器温度]

变频器功率模块的当前工作温度。

参数值	缺省值	只读
	最小值/最大值	0/120℃
	显示单位	1℃

d025[计数器状态]

计数器工作时,它的当前数值。

参数值	缺省值	只读
	最小值/最大值	0/9 999
	显示单位	1

d026[定时器状态]

 32 位参数

定时器使用时,它的当前数值。

参数值	缺省值	只读
	最小值/最大值	0.0/9 999 s
	显示单位	0.1 s

d028[步序逻辑状态]

当参数 P038[速度基准值]设置为 6"步序逻辑"时,该参数将显示由参数 A140—A147 确定的步序逻辑图的当前步序。

参数值	缺省值	只读
	最小值/最大值	0/7
	显示单位	1

d029[转矩电流]
电动机转矩电流的数值。

参数值	缺省值	只读
	最小值/最大值	0.0/变频器额定电流×2
	显示单位	0.01 A

基本程序(设置)组

P031[电动机铭牌电压]　　　　　　　　　　　相关参数：d004,A084,A085,A086,A087
○ 改变参数前,停止变频器运行。
设置电动机铭牌的额定电压。

参数值	缺省值	基于变频器额定值
	最小值/最大值	20/变频器额定电压
	显示单位	1 V AC

P032[电动机铭牌频率]　　　　　　　　　　　相关参数：A084,A085,A086,A087,A090
○ 改变参数前,停止变频器运行。
设置电动机铭牌的额定频率。

参数值	缺省值	60 Hz
	最小值/最大值	15/400 Hz
	显示单位	1 Hz

P033[电动机过载电流]　　　相关参数：A055,A058,A061,A089,A090,A098,A114,A118
设置为电动机最大允许电流。
如果该参数值持续 60 s 超过 150%,则变频器会显示故障 F7 电动机过载故障。

参数值	缺省值	基于变频器额定值
	最小值/最大值	0.0/(变频器额定电流×2)
	显示单位	0.1 A

P034[最小频率] 相关参数:d001,d002,d013,P035,A085,A086,A087,A110,A112

设置变频器所能持续输出的最低频率。

参数值	缺省值	0.0 Hz
	最小值/最大值	0.0/400.0 Hz
	显示单位	0.1 Hz

P035[最大频率] 相关参数:d001,d002,d013,P034,A065,A078,A085,A086,A087,A111,A113

🛇 改变参数前,停止变频器运行。

设置变频器所能连续输出的最高频率。

参数值	缺省值	60 Hz
	最小值/最大值	0/400 Hz
	显示单位	1 Hz

P038[速度基准值]
相关参数:d001,d002,d012,d020,d021,P039,P040,A051—A054,A069,A070—A077,A110,A111,A112,A113,A123,A132,A140—A147,A150—A157

设置变频器速度基准值的来源。

变频器速度命令值有许多不同的来源。来源通常由参数 P038[速度基准值]决定。然而,如果参数 A051—A054[数字量输入×选择]设置为选项 2,4,5,6,11,12,13,14,15 并且数字量输入被激活,或者参数 A132[PID 基准值选择]没有被设置为"0",那么由参数 P038[速度基准值]定义的速度基准值将会被覆盖。

	0	"变频器电位器"(缺省)	内部频率命令,来自集成式键盘上的电位器
选项	1	"内部频率"	内部频率命令,来自参数 A069[内部频率]。当使用 MOP 功能时必须设置
	2	"0~10 V 输入"	外部频率命令,来自 0~10 V 或者±10 V 模拟量输入或者远程电位计
	3	"4~20 mA 输入"	外部频率命令,来自 4~20 mA 模拟量输入
选项	4	"预置频率"	外部频率命令,当参数 A051—A054[数字输入×选择]编程为"预置频率",并且数字量输入被激活时,由参数 A070—A077[预置频率×]决定
	5	"通信端口"	外部频率命令,来自通信端口
	6	"步序逻辑"	外部频率命令,由参数 A070—A077[预置频率×]和 A140—A147[步序逻辑×]决定
	7	"模拟量输入相乘"	外部频率命令,由模拟量输入(由参数 d020[模拟量输入 0~10 V]和 d021[模拟量输入 4~20 mA]显示)的乘积决定。 [模拟量输入 0~10 V]×[模拟量输入 4~20 mA]=速度命令 示例:100%×50%=50%

P039[加速时间 1] 相关参数:P038,P040,A051—A054,A067,A070—A077,A140—A147

设置所有速度增加的加速速率。

最大输出频率/加速时间=加速速率

附录 B PowerFlex 40 变频器常用参数

参数值	缺省值	10.0 s
	最小值/最大值	0.0/600.0 s
	显示单位	0.1 s

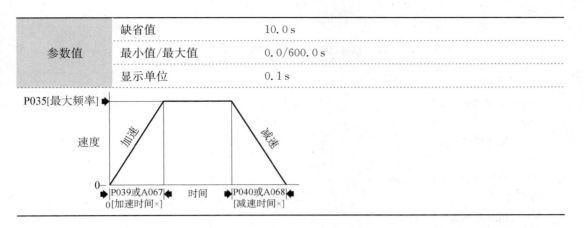

P040[减速时间 1]　　　　相关参数：P038,P039,A051—A054,A068,A070—A077,A140—A147

设置所有速度下降时的减速速率。

最大输出频率/减速时间＝减速速率

参数值	缺省值	10.0 s
	最小值/最大值	0.1/600.0 s
	显示单位	0.1 s

P041[复位成缺省值]

　　改变参数前,停止变频器运行。

将所有参数值复位成出厂缺省值。

选项	0	"预备/空闲"(缺省)	
	1	"出厂复位"	• 在复位功能完成后,该参数将自动重新设置成 0。 • 引起 F48 参数复位成缺省值故障

P042[电压等级]

　　改变参数前,停止变频器运行。

设置变频器的电压等级。

选项	2	"低电压"	480 V
	3	"高电压"(缺省)	600 V

高级编程(设置)组

A051[数字量输入 1 选择]
(I/O 端子 05)

A052[数字量输入 2 选择]
(I/O 端子 06)

A053[数字量输入 3 选择]
(I/O 端子 07)

A054[数字量输入 4 选择]
(I/O 端子 08)

为数字量输入选择功能。

相关参数:d012,d014,P038,P039,P040,A067,A068,
A070—A077,A078,A079,A118,A140—A147

⚫ 改变参数前,停止变频器运行。

选项	0	"未使用"	端子没有任何功能,但是可以由参数 d014[数字量输入状态]通过网络通信读取
	1	"加速和减速 2"	• 当激活时,参数 A067[加速时间 2]和参数 A068[减速时间 2]用于除了点动以外的所有斜坡速率。 • 只能连接 1 个输入
	2	"点动"	• 当输入出现时,变频器根据参数 A079[点动加速/减速]中的设置值进行加速,并且根据 A078[点动频率]中的设置值进行斜坡运行。 • 当输入拿掉时,变频器根据参数 A079[点动加速/减速]中的设置值进行斜坡停止。 • 一个有效的"启动"命令将覆盖这个输入
	3	"辅助故障"	若该选项被使能,当拿掉输入时,将发生 F2 辅助输入故障
	4	"预置频率" (A051 和 A052 的缺省值)	参阅参数 A070—A077[预置频率×]。 **重要事项:**当数字量输入编辑为预置速度并且激活时,它们具有频率控制的优先权
	5	"本地" (A053 的缺省值)	当激活时,数字键盘将成为启动源,而在它上面的电位器将成为速度源
	6	"通信端口"	• 当激活时,通信设备将成为缺省的启动/速度命令源。 • 只能连接 1 个输入
	7	"清除故障"	当激活时,清除一个激活的故障
	8	"斜坡停止,故障清除"	无论参数 P037[停止模式]如何设置,变频器将立刻斜坡停止
	9	"惯性停止,故障清除"	无论参数 P037[停止模式]如何设置,变频器将立刻惯性停止
	10	"直流注入停止,故障清除"	无论参数 P037[停止模式]如何设置,变频器将立刻开始直流注入停止
	11	"点动正向" (A054 缺省值)	变频器根据参数 A079[点动加速/减速]的设置加速到参数 A078[点动频率]的设置值,然后当输入不激活时变频器将斜坡停止。一个有效的启动将会覆盖此命令

续表

选项	12	"点动反向"	变频器根据参数 A079[点动加速/减速]的设置加速到参数 A078[点动频率]的设置值,然后当输入不激活时变频器将斜坡停止。一个有效的启动将会覆盖此命令
	13	"10 V 输入控制"	选择 0～10 V 或者±10 V 控制作为频率基准值。不改变启动源
	14	"20 mA 输入控制"	选择 4～20 mA 控制作为频率基准值。不改变启动源
	15	"PID 禁止"	禁止 PID 功能。变频器使用下一个有效的非 PID 速度基准值
	16	"MOP 增加"	以每秒 2 Hz 的速率增加参数 A069[内部频率]的设置值。A069 的缺省值是 60 Hz
	17	"MOP 减小"	以每秒 2 Hz 的速率减少参数 A069[内部频率]的设置值。A069 的缺省值是 60 Hz
	18	"定时器启动"	清除并且启动定时器功能。可能用于控制继电器或者光电耦合输出
	19	"计数器输入"	启动计数器功能。可能被用于控制继电器或者光电耦合输出
	20	"定时器复位"	清除激活的定时器
	21	"计数器复位"	清除激活的计数器
	22	"定时器和计数器复位"	清除激活的定时器和计数器
	23	"逻辑输入 1"	逻辑功能输入编号 1。可能用于控制继电器或者光电耦合输出(见参数 A055,A058,A061,选项 11～14)。可能与步序逻辑参数 A140—A147[步序逻辑×]联合使用
	24	"逻辑输入 2"	逻辑功能输入编号 2。可能用于控制继电器或者光电耦合输出(见参数 A055,A058,A061,选项 11～14)。可能与步序逻辑参数 A140—A147[步序逻辑×]联合使用
	25	"电流限幅 2"	激活时,参数 A118[电流限制 2]决定变频器的电流限制幅值
	26	"模拟量反向"	通过在参数 A110[模拟量输入 0～10 V 下限]和 A111[模拟量输入 0～10 V 上限]或参数 A112[模拟量输入 4～20 mA 下限]和 A113[模拟量输入 4～20 mA 上限]中设置,将模拟量输入幅值进行反向标定

A055[继电器输出选择]　　　　相关参数:P033,A056,A092,A140—A147,A150—A157,A160,A161
设置改变输出继电器触点状态的条件。

选项	0	"准备好/故障"(缺省值)	变频器上电时继电器改变状态。这表明变频器准备运行。当掉电或者发生故障时,继电器会使变频器返回到闲置状态
	1	"达到频率"	变频器达到命令频率
	2	"电动机运行"	变频器给电动机供电
	3	"反向"	变频器被命令反向运行
	4	"电动机过载"	电动机过载条件存在
	5	"斜坡调节"	斜坡调节器正在调节已编程的加速/减速时间,以避免发生过流或者过压故障

续表

选项	6	"频率超限"	• 变频器超过在参数 A056[继电器输出幅值]中设置的频率值。 • 使用参数 A056 设置极限值
	7	"电流超限"	• 变频器超过在参数 A056[继电器输出幅值]中设置的电流值(%A)。 • 使用参数 A056 设置极限值。 重要事项：参数 A056[继电器输出幅值]的值必须以变频器额定输出电流百分数的形式输入
	8	"直流电压超限"	• 变频器超过参数 A056[继电器输出幅值]中设置的直流母线电压值。 • 使用参数 A056 设置极限值
	9	"退出重试"	超过参数 A092[自动重新启动尝试]中的设置值
	10	"模拟量电压超限"	• 模拟量输入电压(I/O 端子 13)超过参数 A056[继电器输出幅值]中的设置值。 • 当参数 A123[10 V 双极性使能]设置成 1"双极性输入"时不要使用。 • 当输入(I/O 端子 13)接有一个 PTC 和外部电阻时，该参数的设置可以用于表明一个 PTC 跳闸点。 • 使用参数 A056 设置极限值
	11	"逻辑输入 1"	一个输入被编程为"逻辑输入 1"并且被激活
	12	"逻辑输入 2"	一个输入被编程为"逻辑输入 2"并且被激活
	13	"逻辑输入 1 和 2"	两个逻辑输入都被编程并且被激活
	14	"逻辑输入 1 或 2"	一个或者两个逻辑输入被编程并且被激活
	15	"步序逻辑输出"	变频器输入步序逻辑步序，并且命令字(A140—A147)的数字 3 设置成使能步序逻辑输出
	16	"定时器超限"	• 定时器超过了参数 A056[继电器输出幅值]中的设置值。 • 使用参数 A056 设置极限值
	17	"计数器超限"	• 计数器超过了参数 A056[继电器输出幅值]中的设置值。 • 使用参数 A056 设置极限值
	18	"功率因数角超限"	• 功率因数角度超过了参数 A056[继电器输出幅值]中的设置值。 • 使用参数 A056 设置极限值
	19	"模拟量输入丢失"	发生模拟量输入丢失。当发生输入丢失时，编辑参数 A122[模拟量输入丢失]，实现需要的动作
	20	"参数控制"	通过向参数 A056[继电器输出幅值]中赋值，使输出通过网络通信进行控制(0=关(off),1=开(on))
	21	"不可复位的故障"	• 超过了参数 A092[自动重新启动尝试]中设置的数值。 • 参数 A092[自动重新启动尝试]没有使能。 • 发生一个不可复位的故障
	22	"电磁闸制动控制"	给电磁闸制动施加电压。编辑参数 A160[EM 制动关闭(off)延迟]和参数 A161[EM 制动开启(on)延迟]，实现需要的动作

A056[继电器输出幅值]　　　　　　　　　　　　　　　　　相关参数：A055,A058,A061

32 位参数。

如果参数 A055[继电器输出选择]的值是 6,7,8,10,16,17,18 或 20,则为数字量输出继电器设置跳闸点。

附录 B　PowerFlex 40 变频器常用参数　137

A055 设置	A056 最小值/最大值
6	0/400 Hz
7	0/180%
8	0/815 V
10	0/100%
16	0.1/9 999 s
17	1/9 999 个
18	1/180 度
20	0/1

值	缺省值	0
	最小值/最大值	0.0/9 999
	显示单位	0.1

A058[光电耦合输出 1 选择]　　　　　相关参数：P033，A056，A092，A140—A147，A150—A157
A061[光电耦合输出 2 选择]
确定可编程的光电耦合输出的操作。

选项	0	"准备好/故障"	当变频器上电时,光电耦合输出激活。这表明变频器准备运行。当掉电或者发生故障时,光电耦合输出不激活
	1	"达到频率"（A061 缺省值）	变频器达到命令频率
	2	"电动机运行"（A058 缺省值）	变频器给电动机供电
	3	"反向"	变频器被命令反向运行
	4	"电动机过载"	电动机过载条件存在
	5	"斜坡调节"	斜坡调节器正在调节已编程的加速/减速时间,以避免发生过流或者过压故障
	6	"频率超限"	• 变频器超过参数 A059 或 A062[光电耦合输出×幅值]中设置的频率值。 • 使用参数 A059 或/A062 设置极限值
	7	"电流超限"	• 变频器超过参数 A059 或 A062[光电耦合输出×幅值]中设置的电流值(%A)。 • 使用参数 A059 或 A062 设置极限值。 **重要事项**：参数 A059 或 A062[光电耦合输出×幅值]的值必须以变频器额定输出电流百分率的形式输入
	8	"直流电压超限"	• 变频器超过参数 A059 或 A062[光电耦合输出×幅值]中设置的直流母线电压值。 • 使用参数 A059 或 A062 设置极限值

续表

选项	9	"退出重试"	超过参数 A092[自动重新启动尝试]中的设置值
	10	"模拟量电压超限"	• 模拟量输入电压(I/O 端子 13)超过参数 A059 或 A062[光电耦合输出×幅值]中的设置值。 • 当参数 A123[10 V 双极性使能]设置成 1"双极性输入"时不要使用。 • 当输入(I/O 端子 13)接有一个 PTC 和外部电阻器时,该参数的设置可以用于表明一个 PTC 跳闸点。 • 使用参数 A059 或 A062 设置极限值
	11	"逻辑输入 1"	一个输入被编程为"逻辑输入 1"并且被激活
	12	"逻辑输入 2"	一个输入被编程为"逻辑输入 2"并且被激活
	13	"逻辑输入 1 和 2"	两个逻辑输入都被编程并且激活
	14	"逻辑输入 1 或 2"	一个或者两个逻辑输入被编程并且激活
	15	"步序逻辑输出"	变频器输入步序逻辑步序,并且命令字(A140—A147)的数字 3 设置成使能步序逻辑输出
	16	"定时器超限"	• 定时器超过了参数 A059 或 A062[光电耦合输出×幅值]中的设置值。 • 使用参数 A059 或 A062 设置极限值
	17	"计数器超限"	• 计数器达到了参数 A059 或 A062[光电耦合输出×幅值]中的设置值。 • 使用参数 A059 或 A062 设置极限值
	18	"功率因数角超限"	• 功率因数角超过了参数 A059 或 A062[光电耦合输出×幅值]中的设置值。 • 使用参数 A059 或 A062 设置极限值
	19	"模拟量输入丢失"	发生模拟量输入丢失。当发生输入丢失时,编辑参数 A122[模拟量输入丢失],完成需要的动作
	20	"参数控制"	通过向参数 A059 或 A062[光电耦合输出×幅值]中赋值,使输出通过网络通信进行控制(0=关(off),1=开(on))
	21	"不可复位的故障"	• 超过了参数 A092[自动重新启动尝试]中设置的数值。 • 参数 A092[自动重新启动尝试]没有使能。 • 发生一个不可复位的故障
	22	"电磁闸制动控制"	给电磁闸制动施加电压。编辑参数 A160[EM 制动关闭(off)延迟]和参数 A161[EM 制动开启(on)延迟],实现需要的动作

A065[模拟量输出选择]　　　　　　　　　　　　　　　　　　　　　　　相关参数:P035,A066

设置模拟量输出信号模式(0~10 V,0~20 mA,或者 4~20 mA)。输出提供一个信号量与多种可编程信号相匹配。

附录 B　PowerFlex 40 变频器常用参数

选项	输出范围	最小输出值	最大输出值 A066 [模拟量输出上限]	DIP 开关位置	相关参数
0 "输出频率 0～10"	0～10 V	0 V=0 Hz	P035[最大频率]	0～10 V	d001
1 "输出电流 0～10"	0～10 V	0 V=0 A	200%变频器额定输出电流	0～10 V	d003
2 "输出电压 0～10"	0～10 V	0 V=0 V	120%变频器额定输出电压	0～10 V	d004
3 "输出功率 0～10"	0～10 V	0 V=0 kW	200%变频器额定功率	0～10 V	d022
4 "测试数据 0～10"	0～10 V	0 V=0 000	65 535(16 进制 FFFF)	0～10 V	d019
5 "输出频率 0～20"	0～20 mA	0 mA=0 Hz	P035[最大频率]	0～20 mA	d001
6 "输出电流 0～20"	0～20 mA	0 mA=0 A	200%变频器额定输出电流	0～20 mA	d003
7 "输出电压 0～20"	0～20 mA	0 mA=0 V	120%变频器额定输出电压	0～20 mA	d004
8 "输出功率 0～20"	0～20 mA	0 mA=0 kW	200%变频器额定功率	0～20 mA	d022
9 "测试数据 0～20"	0～20 mA	0 mA=0 000	65 535(16 进制 FFFF)	0～20 mA	d019
10 "输出频率 4～20"	4～20 mA	4 mA=0 Hz	P035[最大频率]	0～20 mA	d001
11 "输出电流 4～20"	4～20 mA	4 mA=0 A	200%变频器额定输出电流	0～20 mA	d003
12 "输出电压 4～20"	4～20 mA	4 mA=0 V	120%变频器额定输出电压	0～20 mA	d004
13 "输出功率 4～20"	4～20 mA	4 mA=0 kW	200%变频器额定功率	0～20 mA	d022
14 "测试数据 4～20"	4～20 mA	4 mA=0 000	65 535(16 进制 FFFF)	0～20 mA	d019
15 "输出转矩 0～10"	0～10 V	0 V=0 A	200%变频器额定 FLA	0～10 V	d029
16 "输出转矩 0～20"	0～20 mA	0 mA=0 A	200%变频器额定 FLA	0～20 mA	d029
17 "输出转矩 4～20"	4～20 mA	4 mA=0 A	200%变频器额定 FLA	0～20 mA	d029
18 "设定点 0～10"	0～10 V	0 V=0%	100.0%设定点设置	0～10 V	A109
19 "设定点 0～20"	0～20 mA	0 mA=0%	100.0%设定点设置	0～20 mA	A109
20 "设定点 4～20"	4～20 mA	4 mA=0%	100.0%设定点设置	0～20 mA	A109

值	缺省值	0
	最小值/最大值	0/20
	显示单位	1

A066[模拟量输出上限]　　　　　　　　　　　　　　　　　　　　相关参数：A065

为参数 A065[模拟量输出选择]的源设置标定最大输出值。

示例：

A066 设置	A065 设置	A065 最大输出值
50%	1"输出电流 0～10"	200%变频器额定输出电流时 5 V
90%	8"输出功率 0～20"	200%变频器额定功率时 18 mA

值	缺省值	100%
	最小值/最大值	0/800%
	显示单位	1%

A067[加速时间 2]　　　　　　　　　相关参数：P039，A051—A054，A070—A077，A140—A147

当激活时，为除了点动之外的所有速度增加设置加速速率。

最大频率/加速时间＝加速速率

值	缺省值	20.0 s
	最小值/最大值	0.0/600.0 s
	显示单位	0.1 s

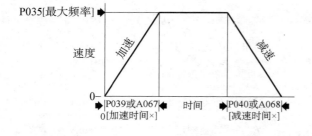

A068[减速时间 2]　　　　　　　　　相关参数：P040，A051—A054，A070—A077，A140—A147

当激活时，为除了点动之外的所有速度减少设置减速速率。

最大频率/减速时间＝减速速率

值	缺省值	20.0 s
	最小值/最大值	0.1/600.0 s
	显示单位	0.1 s

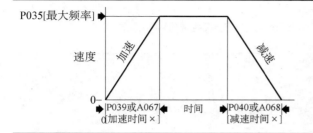

A069[内部频率] 相关参数:P038,A162

当参数 P038[速度基准值]设置成 1"内部频率"时,为变频器提供频率命令。当使能时,该参数在编程模式下,通过使用数字键盘的上下键实时改变频率命令。

重要事项: 一旦达到需要的命令频率,按下 Enter 键将该值保存到 EEPROM 内存中。如果先按下 ESC 键,则频率将沿着通常的加速/减速曲线返回到初值。如果参数 A051—A054[数字量输入×选择]设置成 16"MOP 增加"或者 17"MOP 减小",则该参数作为 MOP 频率基准值。

值	缺省值	60 Hz
	最小值/最大值	0.0/400.0 Hz
	显示单位	0.1 Hz

A070[预置频率 0][1]

相关参数:P038,P039,P040,A051—A053,A067,A068,A140—A147,A150—A157

A071[预置频率 1]
A072[预置频率 2]
A073[预置频率 3]
A074[预置频率 4]
A075[预置频率 5]
A076[预置频率 6]
A077[预置频率 7]

值	A070 缺省值[1]	0.0 Hz
	A071 缺省值	5.0 Hz
	A072 缺省值	10.0 Hz
	A073 缺省值	20.0 Hz
	A074 缺省值	30.0 Hz
	A075 缺省值	40.0 Hz
	A076 缺省值	50.0 Hz
	A077 缺省值	60.0 Hz
	最小值/最大值	0.0/400.0 Hz
	显示单位	0.1 Hz

当参数 A051—A053[数字量输入×选择]设置成选项 4"设置频率"时,提供固定的频率命令值。任何一个激活的预置输入将会覆盖速度命令。

(1) 要激活参数 A070[预置频率 0],将参数 P038[速度基准值]设置成选项 4"预置频率 0～3"。

数字量输入 1 的输入状态(当参数 A051=4 时,I/O 端子 05 的状态)	数字量输入 2 的输入状态(当参数 A052=4 时,I/O 端子 06 的状态)	数字量输入 3 的输入状态(当参数 A053=4 时,I/O 端子 07 的状态)	频率源	使用的加速/减速参数[2]
0	0	0	A070[预置频率 0]	[加速时间 1]/[减速时间 1]
1	0	0	A071[预置频率 1]	[加速时间 1]/[减速时间 1]
0	1	0	A072[预置频率 2]	[加速时间 2]/[减速时间 2]
1	1	0	A073[预置频率 3]	[加速时间 2]/[减速时间 2]
0	0	1	A074[预置频率 4]	[加速时间 1]/[减速时间 1]
1	0	1	A075[预置频率 5]	[加速时间 1]/[减速时间 1]
0	1	1	A076[预置频率 6]	[加速时间 2]/[减速时间 2]
1	1	1	A077[预置频率 7]	[加速时间 2]/[减速时间 2]

(2) 当数字量输入设置成"加速 2 和减速 2",并且被激活时,该输入将覆盖上表中的设置。

A078[点动频率]   相关参数:P035,A051—A054,A079

使用点动命令时,设置输出频率。

值	缺省值	10 Hz
	最小值/最大值	0.0/[最大频率]
	显示单位	0.1 Hz

A079[点动加速/减速]   相关参数:A078,A051—A054

使用点动命令时,设置加速和减速时间。

值	缺省值	10.0 s
	最小值/最大值	0.1/600.0 s
	显示单位	0.1 s

附录 B　PowerFlex 40 变频器常用参数

A080[直流制动时间]　　　　　　　　　　　　　　　　　　　　　　　相关参数：P037，A081

设置直流制动电流"注入"到电动机内的时间长度。参照参数 A081[直流制动幅值]。

值	缺省值	0.0 s
	最小值/最大值	0.0/99.9 s(设置 99.9 为持续的)
	显示单位	0.1 s

A081[直流制动幅值]　　　　　　　　　　　　　　　　　　　　　　　相关参数：P037，A080

当参数 P037[停止模式]设置成"斜坡"或者"直流制动"时，决定提供给电动机的最大直流制动电流，以 A 为单位。

值	缺省值	变频器额定电流×0.05
	最小值/最大值	0.0/(变频器额定电流×1.8)
	显示单位	0.1 A

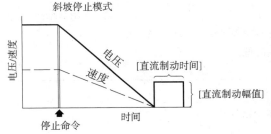
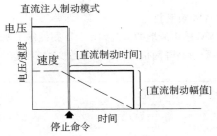

⚠ **注意事项**：如果由于设备或者材料的运动产生伤害的危险，必须使用一个辅助的机械制动设备。
⚠ **注意事项**：该特性不能应用于同步或者永磁电动机。电动机可能在制动过程中被消磁。

A088[最大电压]　　　　　　　　　　　　　　　　　　　　　　相关参数：d004，A085，A086，A087

设置变频器输出的最高电压。

值	缺省值	变频器额定电压
	最小值/最大值	20/变频器额定电压
	显示单位	1 V AC

A089[电流限幅 1]　　　　　　　　　　　　　　　　　　　　　　　相关参数：P033，A118

在电流限幅发生前允许的最大输出电流值。

值	缺省值	变频器额定电流×1.5
	最小值/最大值	0.1/变频器额定电流×1.8
	显示单位	0.1 A

A090 [电动机过载选择]

相关参数：P032, P033

变频器提供了 10 级的电动机过载保护。设置 0～2 来选择 12 t 过载功能的降额系数。

选项	0	"无降额"（缺省值）
	1	"最小降额"
	2	"最大降额"

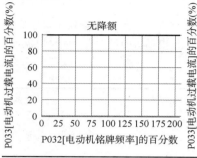 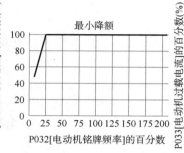 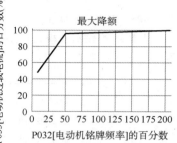

A091 [PWM 频率]

相关参数：A124

设置 PWM 输出波形的载波频率。下图提供了基于 PWM 频率设置的降额指南。

重要事项：忽略降额指南可能会导致变频器性能的降低。

值	缺省值	4.0 kHz
	最小值/最大值	2.0/16.0 kHz
	显示单位	0.1 kHz

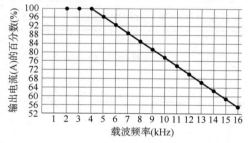

A100 [故障清除]

⏹ 改变此参数前，停止变频器。

复位故障并清除故障队列。主要使用此参数通过网络通信清除故障。

选项	0	"准备好/空闲"（缺省值）
	1	"复位故障"
	2	"清除缓存"　（参数 d007—d009 [故障代码×]）

附录 B　PowerFlex 40 变频器常用参数

A101[编程锁定]
保护参数,以防止被未经许可的人更改。

选项	0	"未锁定"(缺省值)
	1	"锁定"

A102[测试点选择]　　　　　　　　　　　　　　　　　　　　　　　　　相关参数:d019
供罗克韦尔自动化现场技术服务人员使用。

值	缺省值	400
	最小值/最大值	0/FFFF
	显示单位	1(十六进制)

A103[通信数据传输率]　　　　　　　　　　　　　　　　　　　　　　　相关参数:d015
设置 RS-485(DSI)端口的串行口波特率。
重要事项:参数修改后,变频器必须重新上电,使其生效。

选项	0	"1 200"
	1	"2 400"
	2	"4 800"
	3	"9 600"(缺省值)
	4	"19.2 K"
	5	"38.4 K"

A104[通信节点地址]　　　　　　　　　　　　　　　　　　　　　　　相关参数:d015
如果使用网络连接,为 RS-485(DSI)端口设置变频器的节点地址。
重要事项:参数修改后,变频器必须重新上电,使其生效。

值	缺省值	100
	最小值/最大值	1/247
	显示单位	1

A109[模拟量输出设定点]　　　　　　　　　　　　　　　　　　　　　相关参数:A065
当参数 A065[模拟量输出选择]设置成选项 18,19 或 20 时,这个参数设置模拟量输出期望值的百分比。

值	缺省值	0.00%
	最小值/最大值	0.0/100.0%
	显示单位	0.10%

A110[模拟量输入 0～10 V 下限]　　　　　　　　　　　　相关参数:d020,P034,P038,A122

 改变此参数前,停止变频器。

如果参数 P038[速度基准值]使用 0～10 V 输入,将模拟量输入幅值设置成与参数 P034[最小频率]相对应的值。

将该参数设置成比参数 A111[模拟量输入 0～10 V 上限]的值大,可以实现模拟量反向。

值	缺省值	0.00%
	最小值/最大值	0.0/100.0%
	显示单位	0.10%

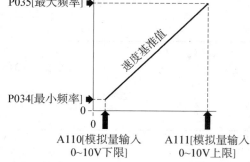

A111[模拟量输入 0～10 V 上限]　　　　　　　　　相关参数:d020,P035,P038,A122,A123

 改变参数前,停止变频器。

如果参数 P038[速度基准值]使用 0～10 V 输入,将模拟量输入幅值设置成与参数 P035[最大频率]相对应的值。

将该参数设置成比参数 A110[模拟量输入 0～10 V 下限]的值小,可以实现模拟量反向。

值	缺省值	100.00%
	最小值/最大值	0.0/100.0%
	显示单位	0.10%

A112[模拟量输入 4～20 mA 下限]　　　　　　　　　　　相关参数:d021,P034,P038

 改变参数前,停止变频器。

如果参数 P038[速度基准值]使用 4～20 mA 输入,将模拟量输入幅值设置成与参数 P034[最小频率]相对应的值。

将该参数设置成比参数 A113[模拟量输入 4～20 mA 上限]的值小,可以实现模拟量反向。

值	缺省值	0.00%
	最小值/最大值	0.0/100.0%
	显示单位	0.10%

A113[模拟量输入 4~20 mA 上限]　　　　　　　　　　　　　　　　　　相关参数:d021,P035,P038

在改变此参数前将变频器停止。

如果参数 P035[速度基准值]使用 4~20 mA 输入,将模拟量输入幅值设置成与参数 P035[最大频率]相对应的值。

将该参数设置成比参数 A112[模拟量输入 4~20 mA 下限]的值小,可以实现模拟量反向。

值	缺省值	100.00%
	最小值/最大值	0.0/100.0%
	显示单位	0.10%

A114[满载滑差频率补偿]　　　　　　　　　　　　　　　　　　　　　　　　　　相关参数:P033

为了补偿感应式电动机固有的滑差,该频率添加到基于电动机电流的命令输出频率上。

值	缺省值	2.0 Hz
	最小值/最大值	0.0/10.0 Hz
	显示单位	0.1 Hz

A115[过程时间下限]　　　　　　　　　　　　　　　　　　　　　　　相关参数:d010,P034

当变频器以参数 P034[最小频率]运行时,该参数用来标定时间值。当设置成非零值时,参数 d010[过程显示]显示过程的持续时间。

值	缺省值	0
	最小值/最大值	0.00/99.9
	显示单位	0.1

A116[过程时间上限]　　　　　　　　　　　　　　　　　　　　　　　相关参数:d010,P035

当变频器以参数 P035[最大频率]运行时,该参数用来标定时间值。当设置成非零值时,参数 d010[过程显示]显示过程的持续时间。

值	缺省值	0
	最小值/最大值	0.00/99.9
	显示单位	0.1

A117[母线调节模式]

禁止母线调节器。

选项	0	"禁止"
	1	"使能"(缺省值)

A118[电流限幅 2]　　　　　　　　　　　　　　　　相关参数：P033，A051—A054，A089

电流限幅发生前允许的最大输出电流值。只有当参数 A051—A054[数字量输入×选择]设置为 25"电流限幅 2"并且激活时,此参数才激活。

值	缺省值	变频器额定电流×1.5
	最小值/最大值	0.1/变频器额定电流×1.8
	显示单位	0.1 A

参考文献

[1] 李华德. 交流调速控制系统[M]. 北京：电子工业出版社，2003.
[2] 钱晓龙. 循序渐进 PowerFlex 变频器[M]. 北京：机械工业出版社，2007.
[3] 孙刚. AB 变频器及其控制技术[M]. 北京：机械工业出版社，2012.
[4] 梁清华，赵越岭，戴永彬，等. 工业控制网络技术实验教程[M]. 沈阳：东北大学出版社，2013.

图书在版编目(CIP)数据

工业变频器技术应用教程/白锐等编著. —上海：复旦大学出版社，2023.6
(复旦卓越. 电工电子系列)
ISBN 978-7-309-16656-9

Ⅰ.①工⋯　Ⅱ.①白⋯　Ⅲ.①变频器-教材　Ⅳ.①TN773

中国版本图书馆 CIP 数据核字(2022)第 237555 号

工业变频器技术应用教程
白　锐　王贺彬　赵越岭　吴　静　编著
责任编辑/李小敏

复旦大学出版社有限公司出版发行
上海市国权路 579 号　邮编：200433
网址：fupnet@ fudanpress.com　http：//www.fudanpress.com
门市零售：86-21-65102580　　团体订购：86-21-65104505
出版部电话：86-21-65642845
上海华业装潢印刷厂有限公司

开本 787×1092　1/16　印张 9.75　字数 237 千
2023 年 6 月第 1 版
2023 年 6 月第 1 版第 1 次印刷

ISBN 978-7-309-16656-9/T·731
定价：29.00 元

如有印装质量问题，请向复旦大学出版社有限公司出版部调换。
版权所有　　侵权必究